Universitext

*Editorial Board
(North America):*

S. Axler
F.W. Gehring
P.R. Halmos

Springer
*New York
Berlin
Heidelberg
Barcelona
Budapest
Hong Kong
London
Milan
Paris
Santa Clara
Singapore
Tokyo*

Universitext

Editors (North America): S. Axler, F.W. Gehring, and P.R. Halmos

Aksoy/Khamsi: Nonstandard Methods in Fixed Point Theory
Aupetit: A Primer on Spectral Theory
Booss/Bleecker: Topology and Analysis
Borkar: Probability Theory: An Advanced Course
Carleson/Gamelin: Complex Dynamics
Cecil: Lie Sphere Geometry: With Applications to Submanifolds
Chae: Lebesgue Integration (2nd ed.)
Charlap: Bieberbach Groups and Flat Manifolds
Chern: Complex Manifolds Without Potential Theory
Cohn: A Classical Invitation to Algebraic Numbers and Class Fields
Curtis: Abstract Linear Algebra
Curtis: Matrix Groups
DiBenedetto: Degenerate Parabolic Equations
Dimca: Singularities and Topology of Hypersurfaces
Edwards: A Formal Background to Mathematics I a/b
Edwards: A Formal Background to Mathematics II a/b
Foulds: Graph Theory Applications
Fuhrmann: A Polynomial Approach to Linear Algebra
Gardiner: A First Course in Group Theory
Gårding/Tambour: Algebra for Computer Science
Goldblatt: Orthogonality and Spacetime Geometry
Gustafson/Rao: Numerical Range: The Field of Values of Linear Operators and Matrices
Hahn: Quadratic Algebras, Clifford Algebras, and Arithmetic Witt Groups
Holmgren: A First Course in Discrete Dynamical Systems
Howe/Tan: Non-Abelian Harmonic Analysis: Applications of $SL(2, R)$
Howes: Modern Analysis and Topology
Humi/Miller: Second Course in Ordinary Differential Equations
Hurwitz/Kritikos: Lectures on Number Theory
Jennings: Modern Geometry with Applications
Jones/Morris/Pearson: Abstract Algebra and Famous Impossibilities
Kannan/Krueger: Advanced Analysis
Kelly/Matthews: The Non-Euclidean Hyperbolic Plane
Kostrikin: Introduction to Algebra
Luecking/Rubel: Complex Analysis: A Functional Analysis Approach
MacLane/Moerdijk: Sheaves in Geometry and Logic
Marcus: Number Fields
McCarthy: Introduction to Arithmetical Functions
Meyer: Essential Mathematics for Applied Fields
Mines/Richman/Ruitenburg: A Course in Constructive Algebra
Moise: Introductory Problems Course in Analysis and Topology
Morris: Introduction to Game Theory
Porter/Woods: Extensions and Absolutes of Hausdorff Spaces
Ramsay/Richtmyer: Introduction to Hyperbolic Geometry
Reisel: Elementary Theory of Metric Spaces
Rickart: Natural Function Algebras
Rotman: Galois Theory

(continued after index)

Karl E. Gustafson   Duggirala K.M. Rao

Numerical Range

The Field of Values of Linear Operators and Matrices

With 9 Figures

Springer

Karl E. Gustafson
Department of Mathematics
University of Colorado at Boulder
Boulder, CO 80309-0395
USA

Duggirala K.M. Rao
Department of Mathematics
Colegio Bolivar at Cali
Cali, Colombia

Editorial Board
(North America):

S. Axler
Department of Mathematics
Michigan State University
East Lansing, MI 48824
USA

F.W. Gehring
Department of Mathematics
University of Michigan
Ann Arbor, MI 48109
USA

P.R. Halmos
Department of Mathematics
Santa Clara University
Santa Clara, CA 95053
USA

Mathematics Subject Classification (1991): 47A12

Library of Congress Cataloging-in-Publication Data
Gustafson, Karl E.
    Numerical range:the field of values of linear operators and
matrices/Karl E. Gustafson, Duggirala K.M. Rao.
      p.   cm. — (Universitext)
    Includes bibliographical references and index.
    ISBN 0-387-94835-X (softcover)
    1. Numerical range. I. Rao, Duggirala K.M. II. Title.
QA329.2.G88 1996
515'.7246—dc20   96-23980

Printed on acid-free paper.

© 1997 Springer-Verlag New York, Inc.
All rights reserved. This work may not be translated or copied in whole or in part without the written permission of the publisher (Springer-Verlag New York, Inc., 175 Fifth Avenue, New York, NY 10010, USA), except for brief excerpts in connection with reviews or scholarly analysis. Use in connection with any form of information storage and retrieval, electronic adaptation, computer software, or by similar or dissimilar methodology now known or hereafter developed is forbidden.
The use of general descriptive names, trade names, trademarks, etc., in this publication, even if the former are not especially identified, is not to be taken as a sign that such names, as understood by the Trade Marks and Merchandise Marks Act, may accordingly be used freely by anyone.

Camera-ready copy prepared from the authors' $\mathcal{A}_{\mathcal{M}}\mathcal{S}$-TEX files.
Printed and bound by Maple-Vail Book Manufacturing Group, York, PA.
Printed in the United States of America.

9 8 7 6 5 4 3 2 1

ISBN 0-387-94835-X Springer-Verlag New York Berlin Heidelberg   SPIN 10543505

To the beautiful interplay
of pure and applied mathematics

Preface

The theories of quadratic forms and their applications appear in many parts of mathematics and the sciences. All students of mathematics have the opportunity to encounter such concepts and applications in their first course in linear algebra. This subject and its extensions to infinite dimensions comprise the theory of the numerical range $W(T)$. There are two competing names for $W(T)$, namely, the *numerical range* of T and the *field of values* for T. The former has been favored historically by the functional analysis community, the latter by the matrix analysis community. It is a toss-up to decide which is preferable, and we have finally chosen the former because it is our habit, it is a more efficient expression, and because in recent conferences dedicated to $W(T)$, even the linear algebra community has adopted it. Also, one universally refers to the numerical radius, and not to the field of values radius. Originally, Toeplitz and Hausdorff called it the *Wertvorrat* of a bilinear form, so other good names would be *value field* or *form values*. The Russian community has referred to it as the *Hausdorff domain*. Murnaghan in his early paper first called it *the region of the complex plane covered by those values* for an $n \times n$ matrix T, then the *range of values* of a Hermitian matrix, then the *field of values* when he analyzed what he called the *sought-for region*. Marshall Stone (1932), in his influential book on operator theory, chose to use the name *numerical range* for $W(T)$.

We know of no book dedicated to presenting the fundamentals of this subject. Our goal here is to do so. Our hope is that this interesting and useful subject will thereby become available to a wider audience. To that end, we have chosen what we call the roadmap approach in writing this book. We want it to be quickly informative as to principal cities and main routes, but without getting the reader lost in secondary byways or overly general description. For this reason, we place and keep the subject squarely in a complex Hilbert space. This setting is the heart of the numerical range theory for bounded linear operators T and naturally contains the field of values theory for finite-dimensional matrices T.

The outline of the book is as follows. In Chapter 1, we have selected the most fundamental properties of the numerical range $W(T)$. These include its convexity and its inclusion of the spectrum of T within its closure. In Chapter 2, we present mapping theorems relating $W(T)$ to the spectral properties of T. The best known of these are probably the power inequality and the dilation theory, but other important mapping properties are given there as well. Chapter 3 describes an operator trigonometry in which we ourselves have played a major role. This includes sharp criteria for accretive operator products $\operatorname{Re} W(T_1 T_2) \geqq 0$, and a theory of antieigenvalues of an arbitrary operator T. We believe the latter will eventually become a standard chapter in linear algebra and will be useful in a variety of applications. Chapter 4 investigates connections between the numerical range $W(T)$ and numerical analysis. This includes applications to schemes used in computational fluid dynamics and an improved convergence theory for certain numerical iterative algorithms from optimization theory. In Chapter 5, we expose important properties of $W(T)$ for matrix theory, the finite-dimensional case. We also develop the essentials of some of the variations of $W(T)$, such as the C numerical range, which have been of recent interest. We conclude in Chapter 6 with a presentation of the properties of certain interesting classes of operators that although no longer symmetric or normal still enjoy some important properties of those operators. These operator classes are defined in terms of their numerical range $W(T)$ properties and include the normaloid, convexoid, and spectraloid operators.

A Word on Notation

We use $T, A, B, \ldots$ to denote a linear operator or matrix, usually bounded and everywhere defined on a finite- or infinite-dimensional Hilbert space $\mathbb{C}^n, H, X, \ldots$. Although the letter T predominates for linear operator, we use whichever letter seems convenient depending on the context or to delineate it from another operator. Recall that T is bounded if the set $\|Tx\|$, where x is in the unit sphere of H (i.e., $\|x\| = 1$), is bounded. In Hilbert space, this is the same as saying that T is continuous. The least upper bound of such $\|Tx\|$ is called the bound M or operator norm $\|T\|$ of T.

In similar fashion, we are not overly worried about forcing a single notation onto the entities that occur in the numerical range literature, which cuts across several mathematical disciplines, most notably functional analysis and linear algebra. Thus, in the literature, the numerical radius may appear as $|W(T)|, w(T)$, or $M(T)$, depending on the context, and the numerical lower bound may appear as $|w(T)|$, when compared to the upper bound $|W(T)|$, or as $m(T)$, or as just m, when compared in the selfadjoint case to the upper bound M. Convexity occurs throughout numerical range research, and one finds notations such as co, Conv, or conv hull for the convex hull of a set. Usually, if it is a toss-up, we have followed the operator theory notation in preference to the matrix theory notations or those notations appearing in inequality theory. For the convenience of the reader, we have included a brief glossary of symbols we frequently use.

We have chosen the bracket notation $\langle y, x \rangle$ for the inner product in the Hilbert space, which has the advantage of no confusion with parentheses in ordinary mathematical or vector expressions. Recall that the inner product is conjugate dual, $\langle y, x \rangle = \overline{\langle x, y \rangle}$, and consistent with the norm, $\langle x, x \rangle = \|x\|^2$. We do not hesitate to identify a linear functional y^* with its representing inner product, i.e., $y^*x = \langle x, y \rangle$ according to the Riesz–Fischer Theorem. In other words, we accept the usual custom of regarding a Hilbert space as selfdual.

A word about the adjoint operator T^*. This is the dual operator, which occurs in some functional analysis contexts as T' or in matrix theory contexts as the conjugate transpose $\bar{T}^T$. Recall that T^* is defined by the duality relationship

$$\langle Tx, y \rangle = \langle x, T^*y \rangle$$

for all x and y in H. The adjoint operator T^* has nothing to do with the adjoint matrix $\mathrm{Adj}(A)$, which occurs as the transpose of the cofactor matrix in an expression for the inverse A^{-1} which occurs in most elementary linear algebra books.

We say that T is selfadjoint when $T = T^*$, which presupposes of course the usual identification of the space H with its dual H^*. This is the same as saying that T is symmetric, if one stays with everywhere defined bounded operators T. In physics, the terminology T is Hermitian is sometimes preferred. In matrix theory, the symmetric matrices $A = A^T$ must have real entries a_{ij}, whereas the larger class of Hermitian matrices $A = \bar{A}^T$ comprises the set of selfadjoint operators. As is well known, selfadjoint operators T have the nice property that all eigenvalues are real. As will be immediately seen, they are exactly those operators whose numerical range $W(T)$ is real.

Prerequisites and References

As prerequisites to this book, the preceding discussion on notation indicates that the reader should have some background experience with the theory of linear operators on a Hilbert space, functional analysis, or linear algebra. Even a first linear algebra course is sufficient to begin the study of the numerical range $W(T)$. We assume that the operator T or matrix T and basis representing it are known to us before we start looking at the properties of T's numerical range $W(T)$. In general, we shall assume that all readers, beginning or expert, are willing to consult their beginning linear algebra or functional analysis textbooks to refresh their memory where needed. There are so many good treatments of those subjects that we would offend all by listing any.

However, for both efficiency and honesty, we would like to mention some references here that provide particularly appropriate background for this book, so that we may refer to them. These include the books that, to our knowledge, present to date the most information on the numerical range

$W(T)$. For functional analysis, one has the classic by F. Riesz and B. Sz. Nagy (1955), *Functional Analysis*, for the theory of linear operators on a Hilbert space developed in a natural way; and for Hilbert space operator theory, P. Halmos (1982), *A Hilbert Space Problem Book*, 2nd ed. Both of these books include a discussion of the numerical range $W(T)$, and the latter has a full chapter on it. For linear algebra, we mention R. A. Horne and C. R. Johnson (1985), *Matrix Analysis*, and its sequel, *Topics in Matrix Analysis* (1991). The latter has a full chapter on the numerical range. For numerical linear algebra, we will refer to G. H. Golub and C. F. Van Loan (1989), *Matrix Computations*, which has recently become a standard treatise. Then there is the interesting recent book, M. Marcus (1993), *Matrices and Matlab*. Not only is the presentation elementary and easy to read, but the author, who has long been interested in $W(T)$, devotes considerable attention to it. For Banach space and Banach algebra generalizations of the numerical range, there are the monographs F. Bonsall and J. Duncan (1971, 1973), *Numerical Ranges of Operators on Normed Spaces and Elements of Normed Algebras* and *Numerical Ranges II*. These monographs are motivated by the application of numerical range concepts to the study of Banach algebras. These background sources will be referred to as follows:

[RN] Riesz and Sz. Nagy, [H] Halmos, [HJ1] Horne and Johnson I, [HJ2] Horne and Johnson II, [GL] Golub and Van Loan, [M] Marcus, [BD] Bonsall and Duncan I, II.

Finally, we appreciate the forebearance of our families during the preparation of this book; the diligent word-processing of the manuscript by Elizabeth Stimmel at UC–Boulder; the assistance of Dr. Guido Sartoris in writing code for the $W(T)$ graphics; the hospitality of the organizers, Professors T. Ando and C. K. Li, and Professor Natalia Bebiano, respectively, at the First and Second Conferences on The Numerical Range and Numerical Radius at the College of William and Mary in Williamsburg, VA, and Universidade de Coimbra, Portugal, in 1992 and 1994, and the Departamento de Matemáticas of the Universidad del Valle, Cali, Colombia, and the Department of Mathematics of the University of Colorado, Boulder, CO, for enabling both of us to visit the other's host institution during this writing.

Karl E. Gustafson
Duggirala K. M. Rao
August, 1995

Glossary

$\mathbb{C}$	the complex numbers
$\mathbb{R}$	the real numbers
$W(T)$	the numerical range of T
$w(T)$	the numerical radius of T
$\partial W(T)$	the boundary of the numerical range
$\overline{W(T)}$	the closure of the numerical range
$\sigma(T)$	the spectrum of T
$\sigma_p(T)$	the point spectrum of T
$\sigma_{\text{app}}(T)$	the approximate point spectrum
$r(T)$	the spectral radius
$\rho(T)$	the resolvent of T
$\|T\|$	the norm of T
$R(T)$	the range of T
$N(T)$	the nullspace of T
$\text{co}(S)$	the convex hull of S
$\text{Int } S$	the interior of S
$\text{sp } M$	the linear span of M
$M \oplus N$	direct sum of M and N
$A \oplus B$	direct sum of A and B
$\tilde{T}$	a dilation of T
$m(T)$	lower numerical bound of T
$M(T)$	upper numerical bound of T
$\phi(T)$	the angle of T
$\cos T$	the cosine of T
$\sin T$	the sine of T
e^T	the exponential of T
$t[u, v]$	sesquilinear form
$\mu(T)$	antieigenvalue of T
$E_A(x)$	quadratic error norm
∇f	gradient of f
$\kappa(T)$	condition number of T
Δt	time step
Δx	grid size
A_d	diffusion matrix
A_c	convection matrix

$\sigma_\epsilon(T)$	pseudo-spectrum of T
tr A	trace of A
M_n	$n \times n$ matrices in $\mathbb{C}^n$
$A \circ B$	Hadamard product
$A \otimes B$	Kronecker product
$V(T)$	spatial numerical range
$W_C(T)$	C-numerical range
$r_C(A)$	C-numerical radius
$W_c(A)$	c-numerical range
$x \prec y$	x majorized by y
$W_k(T)$	k-numerical range
$\tau(T)$	algebraic numerical range
$W_S(T)$	restricted numerical range
$\tau_M(T)$	M-numerical range
$W_\delta(T)$	δ-numerical range
$Z(T)$	symmetric numerical range

Contents

Preface		vii
	A Word on Notation	viii
	Prerequisites and References	ix
Glossary		xi
1	**Numerical Range**	**1**
	Introduction	1
	1.1 Elliptic Range	3
	1.2 Spectral Inclusion	6
	1.3 Numerical Radius	8
	1.4 Normal Operators	15
	1.5 Numerical Boundary	18
	1.6 Other W-Ranges	22
	Endnotes for Chapter 1	24
2	**Mapping Theorems**	**27**
	Introduction	27
	2.1 Radius Mapping	28
	2.2 Analytic Functions	31
	2.3 Rational Functions	32
	2.4 Operator Products	34
	2.5 Commuting Operators	36
	2.6 Dilation Theory	42
	Endnotes for Chapter 2	46
3	**Operator Trigonometry**	**49**
	Introduction	49
	3.1 Operator Angles	49
	3.2 Minmax Equality	53
	3.3 Operator Deviations	56
	3.4 Semigroup Generators	58
	3.5 Accretive Products	61

	3.6	Antieigenvalue Theory	67
		Endnotes for Chapter 3	79
4	**Numerical Analysis**		**80**
		Introduction	80
	4.1	Optimization Algorithms	81
	4.2	Conjugate Gradient	84
	4.3	Discrete Stability	88
	4.4	Fluid Dynamics	93
	4.5	Lax–Wendroff Scheme	98
	4.6	Pseudo Eigenvalues	103
		Endnotes for Chapter 4	107
5	**Finite Dimensions**		**109**
		Introduction	109
	5.1	Value Field	109
	5.2	Gersgorin Sets	113
	5.3	Radius Estimates	118
	5.4	Hadamard Product	123
	5.5	Generalized Ranges	127
	5.6	$W(A)$ Computation	137
		Endnotes for Chapter 5	148
6	**Operator Classes**		**150**
		Introduction	150
	6.1	Resolvent Growth	150
	6.2	Three Classes	153
	6.3	Spectral Sets	159
	6.4	Normality Conditions	162
	6.5	Finite Inclusions	164
	6.6	Beyond Spectraloid	166
		Endnotes for Chapter 6	168
	Bibliography		**171**
	Index		**187**

1
Numerical Range

Introduction

Quadratic forms and their use in linear algebra are quite well known. A natural extension of these ideas in finite- and infinite-dimensional spaces leads us to the numerical range.

We deal with bounded linear operators on a complex, separable Hilbert space H with inner product $\langle\ ,\ \rangle$. Many of these results remain true in the case of real Hilbert space and for nonseparable spaces, but generally we leave it to the reader to check those cases. When specialized to finite-dimensional spaces, the numerical range is often called the field of values.

The *numerical range* of an operator T is the subset of the complex numbers $\mathbb{C}$, given by

$$W(T) = \{\langle Tx, x\rangle,\ x \in H,\ \|x\| = 1\}.$$

The following properties of $W(T)$ are immediate:

$$W(\alpha I + \beta T) = \alpha + \beta W(T) \quad \text{for} \quad \alpha, \beta \in \mathbb{C},$$
$$W(T^*) = \{\bar{\lambda}, \lambda \in W(T)\},$$
$$W(U^*TU) = W(T), \quad \text{for any unitary } U.$$

Example 1. In $\mathbb{C}^2$ let T be the operator defined by the matrix

$$T = \begin{bmatrix} 0 & 1 \\ 0 & 0 \end{bmatrix}.$$

If $x = (f, g)$, $\|x\|^2 = |f|^2 + |g|^2 = 1$, we have $Tx = (g, 0)$ and $\langle Tx, x\rangle = g\bar{f}$. Notice that $|\langle Tx, x\rangle| = |g||f| \leq \frac{1}{2}(|f|^2 + |g|^2) \leq \frac{1}{2}$, and thus

$$W(T) \subset \left\{z : |z| \leq \frac{1}{2}\right\}.$$

Letting $z = re^{i\theta}$, $0 \leq r \leq \frac{1}{2}$, if we choose $x = (\cos\alpha, e^{i\theta}\sin\alpha)$, where $\sin 2\alpha = 2r \leq 1$ and $0 \leq \alpha \leq \frac{\pi}{4}$, we see that

$$\langle Tx, x\rangle = e^{i\theta}\sin\alpha\cos\alpha = re^{i\theta}.$$

Thus $W(T) = \{z : |z| \leq \frac{1}{2}\}$, the full half-disk.

Example 2. Let T be the unilateral shift on H, the Hilbert space ℓ_2 of square summable sequences. For any $f = (f_1, f_2, \ldots) \in H$, $\|f\| = 1$, we have $Tf = (f_2, f_3, \ldots)$ and hence consider

$$\langle Tf, f \rangle = f_1 \bar{f}_2 + f_2 \bar{f}_3 + f_3 \bar{f}_4 + \cdots$$

with $|f_1|^2 + |f_2|^2 + \cdots = 1$. Notice that

$$|\langle Tf, f \rangle| \leq |f_1||f_2| + |f_2||f_3| + \cdots$$
$$\leq \frac{1}{2}[|f_1|^2 + 2|f_2|^2 + 2|f_3|^2 + \cdots] \leq \frac{1}{2}[2 - |f_1|^2].$$

Thus $|\langle Tf, f \rangle| < 1$ if $|f_1| \neq 0$. For $|f_1| = 0$ and f containing a finite number of nonzero entries, we can show in the same way that $|\langle Tf, f \rangle| < 1$ by considering the minimum natural number n for which $f_n \neq 0$.

Thus $W(T)$ is contained in the open unit disk $\{z : |z| < 1\}$. We now show that it is in fact the open unit disk. Let $z = re^{i\theta}$, $0 \leq r < 1$, be any point of this disk. Consider

$$f = (\sqrt{1-r^2}, r\sqrt{1-r^2}\, e^{-i\theta}, r^2\sqrt{1-r^2}\, e^{-2i\theta}, \ldots).$$

Observe that

$$\|f\|^2 = 1 - r^2 + r^2(1-r^2) + r^4(1-r^2) + \cdots = 1.$$

Furthermore,

$$\langle Tf, f \rangle = r(1-r^2)e^{i\theta} + r^3(1-r^2)e^{i\theta} + \cdots$$
$$= re^{i\theta}.$$

The following example shows the calculation of the numerical range as the envelope of a family of circles.

Example 3. Let the transformation $A : \mathbb{C}^2 \to \mathbb{C}^2$ be represented by

$$A = \begin{bmatrix} r & b \\ 0 & -r \end{bmatrix}, \quad r \in \mathbb{R}, \ b \in \mathbb{C}.$$

Let (f, g) be a unit vector in $\mathbb{C}^2$, $f = e^{i\alpha}\cos\theta$, $g = e^{i\beta}\sin\theta$, $\alpha \in \left[0, \frac{\pi}{2}\right]$, $\beta \in [0, 2\pi)$. Then we have

$$Af = (re^{i\alpha}\cos\theta + be^{i\beta}\sin\theta, -re^{i\beta}\sin\theta)$$

and

$$\langle Af, f \rangle = r(\cos^2\theta - \sin^2\theta) + be^{i(\beta-\alpha)}\sin\theta\cos\theta = x + iy,$$
$$x = r\cos 2\theta + \frac{|b|}{2}\sin 2\theta \cos(\beta - \alpha + \gamma),$$
$$y = \frac{|b|}{2}\sin(\beta - \alpha + \gamma)\sin 2\theta, \quad \gamma = \arg b.$$

So
$$(x - r\cos 2\theta)^2 + y^2 = \frac{|b|^2}{4}\sin^2 2\theta.$$

This is a family of circles and we can now obtain their union.
Rewriting this last expression as

$$(x - r\cos \phi)^2 + y^2 = \frac{|b|^2}{4}\sin^2 \phi, \quad 0 \le \phi \le \pi,$$

and differentiating w.r.t. ϕ, we get

$$(x - r\cos \phi)r = \frac{|b|^2}{4}\cos \phi.$$

Eliminating ϕ between the last two equations, one obtains

$$\frac{x^2}{r^2 + (|b|^2/4)} + \frac{y^2}{(|b|^2/4)} = 1.$$

This is an ellipse with center at 0, minor axis b, and major axis $\sqrt{4r^2 + b^2}$.

The most fundamental property of the numerical range is its convexity. Other important properties are that its closure contains the spectrum of the operator and that the numerical radius provides a norm equivalent to the operator norm. These and other basic properties of the numerical range and its boundary are established in this chapter.

1.1 Elliptic Range

Lemma 1.1-1 (Ellipse Lemma). *Let T be an operator on a two-dimensional space. Then $W(T)$ is an ellipse whose foci are the eigenvalues of T.*

Proof. Without loss of generality (see the notes at the end of the section), we can choose T as an upper triangular matrix

(1.1-1) $$T = \begin{bmatrix} \lambda_1 & a \\ 0 & \lambda_2 \end{bmatrix},$$

where λ_1 and λ_2 are the eigenvalues of T.

If $\lambda_1 = \lambda_2 = \lambda$, we have

$$T - \lambda = \begin{bmatrix} 0 & a \\ 0 & 0 \end{bmatrix}, \quad W(T - \lambda) = \left\{ z : |z| \le \frac{|a|}{2} \right\},$$

(see Example 1) and $W(T)$ is a circle with center at λ and radius $\frac{|a|}{2}$.

If $\lambda_1 \ne \lambda_2$ and $a = 0$, we have

$$T = \begin{bmatrix} \lambda_1 & 0 \\ 0 & \lambda_2 \end{bmatrix}.$$

If $x = (f, g)$, $\langle Tx, x\rangle = \lambda_1|f|^2 + \lambda_2|g|^2 = t\lambda_1 + (1-t)\lambda_2$, where $t = |f|^2$ and $|f|^2 + |g|^2 = 1$. So $W(T)$ is the set of convex combinations of λ_1 and λ_2 and is the segment joining them.

If $\lambda_1 \neq \lambda_2$ and $a \neq 0$, we have

(1.1-2)
$$T - \frac{\lambda_1 + \lambda_2}{2} = \begin{bmatrix} \frac{\lambda_1-\lambda_2}{2} & a \\ 0 & \frac{\lambda_2-\lambda_1}{2} \end{bmatrix},$$
$$e^{-i\theta}\left[T - \frac{\lambda_1 + \lambda_2}{2}\right] = \begin{bmatrix} r & ae^{-i\theta} \\ 0 & -r \end{bmatrix} = B,$$

where $\frac{\lambda_1-\lambda_2}{2} = re^{i\theta}$. $W(B)$ is an ellipse with center at $(0,0)$, and minor axis $|a|$ (see Example 3), and foci at $(r,0)$ and $(-r,0)$. Thus $W(T)$ is an ellipse with foci at λ_1, λ_2, and the major axis has an inclination of θ with the real axis. □

Theorem 1.1-2 (Toeplitz–Hausdorff). *The numerical range of an operator is convex.*

Proof. Let $\alpha, \beta \in W(T)$, $\alpha = \langle Tf, f\rangle$, and $\beta = \langle Tg, g\rangle$, with $\|f\| = \|g\| = 1$. We have to show that the segment containing α and β is contained in $W(T)$. Let V be the subspace spanned by f and g and E the orthogonal projection of H on V, so that $Ef = f$ and $Eg = g$. We also have, for the operator ETE on V,

$$\langle ETEf, f\rangle = \langle Tf, f\rangle$$

and

$$\langle ETEg, g\rangle = \langle Tg, g\rangle.$$

By the ellipse lemma, $W(ETE)$ is an ellipse. Hence $W(ETE)$ contains the segment joining α and β. It is easy to see that $W(ETE) \subset W(T)$ and that $W(T)$ contains the segment joining α and β. □

From Lemma 1.1-1 and Theorem 1.1-2, we see that we may regard the numerical range $W(T)$ as the union of all of its two-dimensional numerical ranges, which are ellipses. These ellipses may be degenerate, as is the case, for example, in the important instance when T is selfadjoint. Then $W(T)$ is just an interval in the real line $\mathbb{R}$. Thus, we see that $W(T)$ need not possess an interior. The ellipse degeneracies also permit straight line polygonal boundaries of $W(T)$. Generally, $W(T)$ is neither open nor closed, except in the finite-dimensional case. Then, it is the jointly continuous image of two compact sets, it is compact.

Notes and References for Section 1.1

The Schur decomposition theorem (see [HJ1, GL, or M]), guarantees that any square matrix T may be transformed by unitary similarity transformation to upper triangular form, with its eigenvalues on the diagonal. Also, since $W(T)$ is invariant under unitary transformations, it suffices to consider only upper triangular matrices in Lemma 1.1-1. Although obvious, this unitary invariance of $W(T)$, as for the spectrum $\sigma(T)$, is essential to many conclusions.

Another proof of Lemma 1.1-1 can be found in
W. F. Donoghue (1957), "On the Numerical Range of a Bounded Operator," *Michigan Math. J.* **4**, 261–263.
An earlier proof was given in the case of finite dimensions in
F. D. Murnaghan (1932). "On the Field of Values of a Square Matrix," *Proc. Nat. Acad. Sci. U.S.A.* **18**, 246–248.

One may also obtain the ellipse nature of $W(A)$ for A a 2×2 matrix by inequalities. By the unitary invariance and by a translation and then a constant normalization, we may assume that

$$A = \begin{bmatrix} r & 2 \\ 0 & -r \end{bmatrix}.$$

Any unit vector f may be written $f = e^{i\psi}(\cos\theta, e^{i\phi}\sin\theta)$. Then

$$f^* A f = (r\cos 2\theta + \cos\phi \sin 2\theta) + i(\sin\phi \sin 2\theta) \equiv x + iy,$$

from which

$$|f^* A f - r\cos 2\theta|^2 = \sin^2 2\theta = (x - r\cos 2\theta)^2 + y^2,$$

which clearly describes a circle (ϕ varying) centered at $r\cos 2\theta$, as in the above proof. Now, instead of differentiating to find the envelope of these circles, we may use the fact that $\cos 2\theta$ must be real and bounded in magnitude by 1. From $(r^2 + 1)\cos^2 2\theta - 2xr\cos 2\theta + (x^2 + y^2 - 1) = 0$, by the quadratic formula we have

$$\cos 2\theta = \frac{xr \pm ((r^2 + 1) - (x^2 + y^2(1 + r^2)))^{\frac{1}{2}}}{r^2 + 1}.$$

That $\cos 2\theta$ must remain real yields the inequality

$$\frac{x^2}{r^2 + 1} + y^2 \leq 1,$$

an ellipse with semiaxes $r^2 + 1$ and 1. Then fixing ϕ and letting θ run shows how the ellipse is filled, in an interesting way: for example, choose $\phi = \pi/2$, then

$$\frac{x^2}{r^2 + 1} + y^2 = \frac{r^2 + \sin^2 2\theta}{r^2 + 1},$$

which varies from 1 down to $r^2/r^2 + 1$. This is a shell-like filling action. One can, of course, by controlling the way in which ϕ and θ run, follow other filling recipes. The envelope-finding proof we gave in Lemma 1.1 guarantees that all of these recipes will fill the ellipse.

Originally, Toeplitz proved that the boundary $\partial W(T)$ was convex and later Hausdorff proved that $W(T)$ was simply connected.

O. Toeplitz (1918). "Das algebraische Analogon zu einem Satz von Fejér," *Math. Z.* **2**, 187–197.

F. Hausdorff (1919), "Der Wertvorrat einer Bilinearform," *Math. Z.* **3**, 314–316.

The Toeplitz–Hausdorff Theorem has many proofs. To the best of our knowledge, the most recent is due to

C. K. Li (1994). "C-Numerical Ranges and C-Numerical Radii," *Linear and Multilinear Algebra* **37**, 51–82.

Other references to this theorem can be found in [H, HJ2, M, BD] and M. Goldberg and E. G. Straus (1979). "Norm Properties of C-Numerical Radii," *Linear Alg. Appl.* **24**, 113–131.

A short proof covering unbounded operators was given in

K. E. Gustafson (1970). "The Toeplitz–Hausdorff Theorem for Linear Operators," *Proc. Amer. Math. Soc.* **25**, 203–204.

1.2 Spectral Inclusion

One important use of $W(T)$ is to bound the spectrum $\sigma(T)$. The *spectrum* of an operator T consists of those complex numbers λ such that $T - \lambda I$ is not invertible. For our purpose of showing that the spectrum of an operator is included in its numerical range, it is enough to look at the boundary of the spectrum.

It is well known (see, for example, [H], Problem 63) that the boundary of the spectrum is contained in the *approximate point spectrum* σ_{app}, which consists of complex numbers λ for which there exists a sequence of unit vectors $\{f_n\}$ with $\|(T - \lambda I)f_n\| \to 0$. Since $W(T)$ is convex, it suffices to show that $\sigma_{\text{app}}(T) \subset W(T)$.

Theorem 1.2-1 (Spectral inclusion). *The spectrum of an operator is contained in the closure of its numerical range.*

Proof. Consider any $\lambda \in \sigma_{\text{app}}(T)$ and a sequence $\{f_n\}$ of unit vectors with
$$\|(T - \lambda I)f_n\| \to 0.$$
By the Schwarz inequality,
$$|\langle (T - \lambda I)f_n, f_n \rangle| \le \|(T - \lambda I)f_n\| \to 0.$$
Thus $\langle Tf_n, f_n \rangle \to \lambda$. So $\lambda \in \overline{W(T)}$. □

Notice that the spectral inclusion enables us to locate the spectrum of the sum of any two operators A and B. Even though $\sigma(A+B)$ has nothing to do in general with $\sigma(A)$ and $\sigma(B)$, we still have
$$\sigma(A + B) \subset W(A + B) \subset W(A) + W(B).$$

As the set $\overline{W(T)}$ is convex, we can also conclude that the convex hull $\mathrm{co}\,(\sigma(T))$ is contained in $\overline{W(T)}$. Generally speaking, even though $W(T)$ is often used to bound $\sigma(T)$, the latter can be much smaller.

Example. Let $H = \mathbb{C} \times \mathbb{C}$ and T, the operator represented by a matrix
$$T = \begin{bmatrix} 0 & 0 \\ 1 & 0 \end{bmatrix}.$$
If $f = (f_1, f_2)$, we have $Tf = (0, f_1)$, $\langle Tf, f \rangle = f_1 \bar{f_2}$, and $W(T) = \{\lambda \in \mathbb{C},\ |\lambda| \leq \frac{1}{2}\}$. However, $\sigma(T) = \{0\}$.

By contrast, selfadjoint operators T have their spectra bounded sharply by $W(T)$, as is evidenced by the following simple relations between $\sigma(T)$ and $W(T)$.

Theorem 1.2-2. *T is selfadjoint iff $W(T)$ is real.*

Proof. If T is selfadjoint, we have, for all $f \in H$, $\langle Tf, f \rangle = \langle f, Tf \rangle = \overline{\langle Tf, f \rangle}$, and hence $W(T)$ is real. Conversely, if $\langle Tf, f \rangle$ is real for all $f \in H$, we have $\langle Tf, f \rangle - \langle f, Tf \rangle = 0 = \langle (T - T^*)f, f \rangle$. Thus the operator $T - T^*$ has only $\{0\}$ in its numerical range. As will be shown in the next section, such an operator has to be the null operator. So $T - T^* = 0$ and $T = T^*$. □

Theorem 1.2-3. *Let T be selfadjoint and $W(T)$ = the real interval $[m, M]$. Then $\|T\| = \sup\{|m|, |M|\}$.*

Proof. Let $w(T) = \sup\{|m|, |M|\}$. Then, for any real real $\lambda \neq 0$, we have by the identity

(1.2-1)
$$\begin{aligned} 4\|Tx\|^2 &= \langle T(\lambda x + \lambda^{-1}Tx), \lambda x + \lambda^{-1}Tx \rangle - \langle T(\lambda x - \lambda^{-1}Tx),\\ & \quad \lambda x - \lambda^{-1}Tx \rangle \\ &\leq w(T)[\|\lambda x + \lambda^{-1}Tx\|^2 + \|\lambda x - \lambda^{-1}Tx\|^2 \\ &= 2w(T)(\lambda^2 \|x\|^2 + \lambda^{-2}\|Tx\|^2). \end{aligned}$$

Taking $\lambda^2 = \frac{\|Tx\|}{\|x\|}$ yields $\|Tx\| \leq w(T)\|x\|$. □

Theorem 1.2-4. *Let $\overline{W(T)} = [m, M]$. Then $m, M \in \sigma(T)$.*

Proof. Since $m \in \overline{W(T)}$, there is a sequence of unit vectors $\{f_n\}$ such that $\langle Tf_n, f_n \rangle \to m$. Hence $\|\langle (T-m)f_n, f_n \rangle\| = \|(T-m)^{\frac{1}{2}}f_n\|^2 \to 0$. Also, $\|(T-m)f_n\| \to 0$ and so $m \in \sigma_{\mathrm{app}}(T) \subset \sigma(T)$. □

Notes and References for Section 1.2

The spectral inclusion was first given by

A. Wintner (1929). "Zur Theorie der beschrankten Bilinearformen," *Math. Z.* **30**, 228–282.

The spectral theory and spectral mapping theorem as well as the spectral inclusion can be found in [H]. Let us recall here that those λ not in the spectrum $\sigma(T)$ are called the *resolvent set* $\rho(T)$ of T, and thereupon, the operator $(T - \lambda I)^{-1}$ is called the *resolvent operator* for T.

1.3 Numerical Radius

The *numerical radius* $w(T)$ of an operator T on H is given by
$$w(T) = \sup\{|\lambda|,\ \lambda \in W(T)\}.$$
Notice that, for any vector $x \in H$, we have
$$|\langle Tx, x\rangle| \leq w(T)\|x\|^2.$$

Example. Let T be the (right) shift operator on $\mathbb{C}^n$ defined by
$$T = \begin{bmatrix} 0 & 0 & 0 & & & \\ 1 & 0 & 0 & & \cdots & \\ 0 & 1 & 0 & & & \\ & \cdots & & & & 0 \\ 0 & 0 & 0 & \cdots & 0 & 1 & 0 \end{bmatrix}.$$

If $f \in \mathbb{C}^n$, $f = (f_1, f_2, \ldots, f_n)$, we have $Tf = (0, f_1, f_2, \ldots, f_{n-1})$ and
$$\langle Tf, f\rangle = f_1\bar{f}_2 + f_2\bar{f}_3 + \cdots + f_{n-1}\bar{f}_n.$$
Thus $|\langle Tf, f\rangle| \leq |f_1||f_2| + \cdots + |f_{n-1}||f_n|$. We have thus to calculate
$$\sup\{|f_1||f_2| + \cdots + |f_{n-1}||f_n|\}$$
subject to the condition $\sum_{i=1}^n |f_i|^2 = 1$.

Let $r_i = |f_i|$, and consider the Lagrange function

(1.3-1) $\quad F(r_1, r_2, \ldots, r_n, \lambda) = r_1 r_2 + \cdots + r_{n-1} r_n - \lambda\left(\sum_1^n r_i^2 - 1\right).$

From (1.3-1) we may set

(1.3-2)
$$\frac{\partial F}{\partial r_1} = r_2 - 2\lambda r_1 = 0,$$
$$\frac{\partial F}{\partial r_2} = r_1 + r_3 - 2\lambda r_2 = 0,$$
$$\cdots$$
$$\frac{\partial F}{\partial r_{n-1}} = r_{n-2} + r_n - 2\lambda r_{n-1} = 0,$$
$$\frac{\partial F}{\partial r_n} = r_{n-1} - 2\lambda r_n = 0.$$

Writing $x = (r_1, r_2, \ldots, r_n)$ and

$$A = \begin{bmatrix} 0 & 1 & 0 & \cdots & 0 & 0 \\ 1 & 0 & & & 0 & \\ 0 & 1 & 0 & 1 & \vdots & \vdots \\ & & & & 0 & 1 \\ 0 & 0 & \cdots & & 1 & 0 \end{bmatrix},$$

we can write the equations (1.3-2) in the form

$$Ax = 2\lambda x.$$

Notice that $A = 2B$, where B is the Hermitian matrix which has $\frac{1}{2}$ in the subdiagonal and the superdiagonal and 0 elsewhere. Thus λ is an eigenvalue of B. These eigenvalues of B are known to be (see Notes and References)

$$\lambda = \cos \frac{k\pi}{n+1}, \quad k = 1, 2, \ldots, n.$$

Notice further that by multiplying (1.3-2) by $r_1, r_2, \ldots, r_n$ in order and adding, we get

(1.3-3) $\quad 2(r_1 r_2 + \cdots + r_{n-1} r_n) - 2\lambda(r_1^2 + \cdots + r_n^2) = 0.$

Thus $\lambda = r_1 r_2 + \cdots + r_{n-1} r_n$ since $r_1^2 + r_2^2 + \cdots + r_n^2 = 1$. Since the maximum value of the expression for λ is $w(T)$, we have necessarily that

$$w(T) = \sup \left\{ \cos \frac{k\pi}{n+1}, \quad k = 1, 2, \ldots, n \right\}.$$

Thus

(1.3-4) $\quad w(T) = \cos \dfrac{\pi}{n+1}.$

Notice that as $n \to \infty$, $w(T) \to 1$.

We now show that $w(T)$ is an equivalent norm of T.

Theorem 1.3-1 (Equivalent norm). $w(T) \leq \|T\| \leq 2w(T)$.

Proof. If $\lambda = \langle Tx, x \rangle$ with $\|x\| = 1$, we have by the Schwarz inequality

$$|\lambda| \leq |\langle Tx, x \rangle| \leq \|Tx\| \leq \|T\|.$$

To prove the other inequality, we use the following identity (polarization principle), which may be verified by direct computation,

(1.3-5) $\quad \begin{aligned} 4\langle Tx, y \rangle &= \langle T(x+y), x+y \rangle - \langle T(x-y), x-y \rangle \\ &\quad + i\langle T(x+iy), x+iy \rangle - i\langle T(x-iy), x-iy \rangle. \end{aligned}$

Hence

$$\begin{aligned} 4|\langle Tx, y \rangle| &\leq w(T)[\|x+y\|^2 + \|x-y\|^2 + \|x+iy\|^2 + \|x-iy\|^2] \\ &= 4w(T)[\|x\|^2 + \|y\|^2]. \end{aligned}$$

Choosing $\|x\| = \|y\| = 1$, we have

$$4|\langle Tx, y\rangle| \leq 8w(T),$$

which implies

$$\|T\| \leq 2w(T). \quad \square$$

Theorem 1.3-1 implies that $T = 0$ whenever $w(T) = 0$. Notice that this result is not valid in a real Hilbert space, as the following example shows.

Example. Let $H = R \times R$ and

$$T = \begin{bmatrix} 0 & -1 \\ 1 & 0 \end{bmatrix}.$$

For $f = (f_1, f_2)$, $\|f\| = 1$, we have $Tf = (-f_2, f_1)$ and

$$\langle Tf, f\rangle = 0.$$

However, $\|T\| = 1$.

Let us now look at two extreme cases of the inequality in Theorem 1.3-1. In the following, we refer to the *spectral radius* $r(T) = \sup\{|\lambda|, \lambda \in \sigma(T)\}$ and the *point spectrum* $\sigma_p(T) = \{\lambda \in \sigma(T), Tf = \lambda f \text{ for some } f \in H\}$.

Theorem 1.3-2. If $w(T) = \|T\|$, then

$$r(T) = \|T\|.$$

Proof. Let $w(T) = \|T\| = 1$. Then there is a sequence of unit vectors $\{f_n\}$ such that $\langle Tf_n, f_n\rangle \to \lambda \in W(T)$, $|\lambda| = 1$. From the inequality

$$|\langle Tf_n, f_n\rangle| \leq \|Tf_n\| \leq 1,$$

we have $\|Tf_n\| \to 1$. Hence

$$(1.3\text{-}6) \quad \|(T - \lambda I)f_n\|^2 = \|Tf_n\|^2 - \langle Tf_n, \lambda f_n\rangle - \langle \lambda f_n, Tf_n\rangle + \|f_n\|^2 \to 0.$$

Hence $\lambda \in \sigma_{\text{app}}(T)$ and $r(T) = 1$. $\square$

Theorem 1.3-3. If $\lambda \in W(T)$, $|\lambda| = \|T\|$, then $\lambda \in \sigma_p(T)$.

Proof. Let $\lambda = \langle Tf, f\rangle$, $\|f\| = 1$. Then

$$\|T\| = |\lambda| = |\langle Tf, f\rangle| \leq \|Tf\| \leq \|T\|.$$

So $|\langle Tf, f\rangle| = \|Tf\|\|f\|$. Thus $Tf = \mu f$ for some $\mu \in \mathbb{C}$. However, $\lambda = \langle Tf, f\rangle = \langle \mu f, f\rangle = \mu$ and hence $Tf = \lambda f$. $\square$

Let us now look at the other extreme condition, $w(T) = \frac{1}{2}\|T\|$. A sufficient condition for this in terms of $R(T) = \{Tf, f \in H\}$ and $R(T^*) = \{T^*f, f \in H\}$ is given by the following.

Theorem 1.3-4. If $R(T) \perp R(T^*)$, then $w(T) = \frac{1}{2}\|T\|$.

Proof. Let $f \in H$, $\|f\| = 1$. We can write $f = f_1 + f_2$, where $f_1 \in N(T)$, the null space of T, and $f_2 \in \overline{R(T^*)}$. Recall that $\overline{R(T^*)}^\perp = N(T)$ by the fundamental theorem of linear algebra. Thus we have

$$\langle Tf, f \rangle = \langle T(f_1 + f_2), f_1 + f_2 \rangle = \langle Tf_2, f_1 \rangle$$

since $Tf_1 = 0$ and $\langle Tf_2, f_2 \rangle = \langle f_2, T^*f_2 \rangle = 0$. Thus

(1.3-7) $\qquad |\langle Tf, f \rangle| \leq \|T\|\|f_2\|\|f_1\| \leq \frac{\|T\|}{2}[\|f_1\|^2 + \|f_2\|^2] = \frac{\|T\|}{2}.$

Since f is arbitrary, we have

$$w(T) \leq \frac{\|T\|}{2} \leq w(T). \quad \square$$

We have seen one example of an operator T with $w(T) = \frac{1}{2}\|T\|$, namely the two-dimensional shift

$$S_2 = \begin{bmatrix} 0 & 0 \\ 1 & 0 \end{bmatrix}.$$

The following theorem shows that some operators T with $w(T) = \frac{1}{2}\|T\|$ have S_2 as a component.

Theorem 1.3-5. If $w(T) = \frac{1}{2}\|T\|$ and T attains its norm, then T has a two-dimensional reducing subspace on which it is the shift S_2.

Proof. Let $\|T\| = 1 = 2w(T) = \|Tf_1\| = \|f_1\|$ for some f_1. If $f_2 = Tf_1$, we have $\|f_2\|^2 = \langle f_2, f_2 \rangle = \langle f_2, Tf_1 \rangle = \langle T^*f_2, f_1 \rangle = \|Tf_1\|^2 = \|f_1\|^2$. Hence, as $\|T^*\| = 1$, we have $T^*f_2 = f_1$. Since, for any θ, $w(T) = w(e^{i\theta}T) = \frac{1}{2}$, we have $w(e^{i\theta}T + e^{-i\theta}T^*) \leq 1$. However, the last operator is selfadjoint and hence, by Theorem 1.2-3, we have $\|e^{i\theta}T + e^{-i\theta}T^*\| \leq 1$. In particular, $\|e^{i\theta}Tf_2 + e^{-i\theta}T^*f_2\| \leq 1$. Using $T^*f_2 = f_1$, we have $\|Tf_2\|^2 \leq 2\text{Re}\,[e^{2i\theta}\langle T^2f_2, f_2\rangle]$ for all θ. Hence $Tf_2 = 0$. Similarly, the other inequality $\|e^{i\theta}Tf_1 + e^{-i\theta}T^*f_1\| \leq 1$ yields $T^*f_1 = 0$. We then have $\langle f_1, f_2 \rangle = \langle f_1, Tf_1 \rangle = \langle T^*f_1, f_1 \rangle = 0$. So, f_1, f_2 form an orthonormal basis for a two-dimensional subspace M. From the relations $Tf_1 = f_2$, $T^*f_1 = 0$, $Tf_2 = 0$, and $T^*f_2 = f_1$, we find that M is a reducing subspace for T and that the matrix of T on M is

$$\begin{bmatrix} 0 & 0 \\ 1 & 0 \end{bmatrix}. \quad \square$$

The numerical radius, in addition to being an equivalent norm, also bounds $\|T^n f\|$ for a given f. There has been a considerable amount of research on finding the best constant C such that $\|T^n f\| \leq C\|f\|$ when $w(T) \leq 1$.

12   1. Numerical Range

Theorem 1.3-6. Let $w(T) = 1$. For any f, $\|f\| = 1$, $\|T^n f\| \to \ell \leq 2$. If $\ell = \sqrt{2}$, then $\|T^n x\| = \sqrt{2}$ for $n = 1, 2, \ldots$.

Proof. We construct a sequence of $n+1$ by $n+1$ determinants D_n, choosing $D_0 = 1$ and

(1.3-8) $$D_n = \begin{vmatrix} 1 & -\frac{\|Tf\|^2}{2} & \cdots & & \cdots \\ -\frac{\|Tf\|^2}{2} & \|Tf\|^2 & \cdots & & \\ \vdots & & & & -\frac{\|T^n f\|^2}{2} \\ \cdots & & \cdots & -\frac{\|T^n f\|^2}{2} & \|T^n f\|^2 \end{vmatrix}.$$

We will show that, for all $n = 1, 2, \ldots$, $D_n \geq 0$, and that the sequence $\frac{D_n}{D_{n-1}}$ is decreasing. Choosing

$$y = a_0 x + a_1 T x + \cdots + a_n T^n x,$$

we have, since $w(T) \leq 1$,

$$|\langle Ty, y \rangle| \leq \langle y, y \rangle$$

and hence

(1.3-9) $$\left| a_0 a_1 \|Tx\|^2 + a_1 a_2 \|T^2 x\|^2 + \cdots + \sum_{\substack{i,j \\ i+1 \neq j}}^{n} a_i a_j \langle T^i x, T^j x \rangle \right|$$

$$\leq a_0^2 + a_1^2 \|Tx\|^2 + \cdots + \sum_{\substack{i,j=0 \\ i \neq j}}^{n} a_i a_j \langle T^i x, T^j x \rangle.$$

The cross terms in (1.3-9) can be eliminated by replacing T by $e^{i\theta} T$ and integrating between 0 and 2π. We thus get the inequality

(1.3-10) $a_0 a_1 \|Tx\|^2 + \cdots + a_{n-1} a_n \|T^n x\|^2 \leq a_0^2 + a_1^2 \|Tx\|^2 + \cdots + a_n^2 \|T^n x\|^2$.

This shows that the quadratic form (the right hand side of (1.3-10) minus the left hand side of (1.3-10)) associated with each D_n is nonnegative. This fact is now used as follows. Assuming that $\|Tx\| \neq 0$, we have $D_1 = \|Tx\|^2 - \frac{\|Tx\|^2}{4} > 0$. Consider

(1.3-11) $$D_n = D_{n-1}\|T^n x\| - \frac{1}{4} D_{n-2}\|T^n x\|^2.$$

Let $D_k = 0$ for some $k > 1$. Then $D_{k+1} = -\frac{1}{4}\|T^{k+1} x\|^2 D_{k-1}$, which is impossible unless $\|T^{k+1} x\|^2 = 0$. Hence $\|T^n x\| = 0$ for all $n \geq k$. In this case, $\ell = 0$.

Let us now suppose that $D_n > 0$ for all n. Then

(1.3-12) $$\frac{D_n}{D_{n-1}} = \frac{D_{n-1}\|T^n x\|^2 - \frac{1}{4} D_{n-2}\|T^{n-1} x\|^2}{D_{n-1}} \leq \frac{D_{n-1}}{D_{n-2}},$$

1.3 Numerical Radius   13

so the sequence $\frac{D_n}{D_{n-1}}$ is convergent. Let L be its limit. Solving for $\|T^n x\|$ in (1.3-12) gives

(1.3-13) $\quad \|T^n x\|^2 = \left\{ \frac{D_{n-1}}{D_{n-2}} \pm \left[\frac{D_{n-1}}{D_{n-2}} \right]^2 - \frac{D_n}{D_{n-2}} \right\}^{\frac{1}{2}} \to 2L.$

Since we know that

(1.3-14) $\quad L \leq \frac{D_1}{D_0} = \|Tx\| - \frac{\|Tx\|^2}{4} = 1 - \frac{\|Tx - 1\|^2}{2} \leq 1,$

we may conclude that $\|T^n x\|^2 \to 2L \leq 2$. □

Theorem 1.3-7. *Let $w(T) = 1$ and $\|T^n f\| = 2$ for some n and unit vector f. Then $\|T^k x\| = \sqrt{2}$ for $k = 1, 2, \ldots, n-1$ and $T^{n+1} x = 0$.*

Proof. With the notation of Theorem 1.3-6, we have $\|T^n x\| = 2$. We then have

(1.3-15) $\quad 1 \leq \frac{D_{n-1}}{D_{n-2}} \leq \frac{D_{n-2}}{D_{n-3}} \leq \cdots \leq \frac{D_1}{D_0} \leq 1.$

So $D_{n-1} = D_{n-2} = \cdots = D_1$ and $\|T^k x\|^2 = 2$ for $k = 1, 2, \ldots, n-1$. Also, $D_n = 0$ and

$$D_{n+1} = \|T^{n+1} x\|^2 D_n - \frac{1}{4} \|T^{n+1} x\|^2 D_{n-1} \geq 0,$$

and so $\|T^{n+1} x\| = 0$. □

The following theorem gives a sufficient condition for an operator to be a projection.

Theorem 1.3-8. *If T is idempotent and $w(T) \leq 1$, then T is an orthogonal projection.*

Proof. It is sufficient to prove that $T = 0$ on $R(T)^\perp$. Let $x \in R(T)$ and $y = Tx$. Then for $t > 0$ we have

$$T(x + ty) = Tx + tT^2 x = y + tTx = y + ty.$$

Thus

(1.3-16) $\quad \begin{aligned} \langle T(x + ty), x + ty \rangle &= \langle (1 + t)y, x + ty \rangle \\ &= \langle (1 + t)y, ty \rangle = (1 + t)t\|y\|^2 \end{aligned}$

as $x \perp y$. On the other hand, we have $\langle T(x + ty), x + ty \rangle \leq \|x + ty\|^2$ as $w(T) \leq 1$. We thus have

$$(1 + t)t\|y\|^2 \leq \|x + ty\|^2 = \|x\|^2 + t^2 \|y\|^2,$$

so $t\|y\|^2 \leq \|x\|^2$. Since t is arbitrary, we conclude that $\|y\| = 0$. Therefore, $T = 0$ on $R(T)^\perp$. □

Notes and References for Section 1.3

The example of $w(T)$ for the right shift operator in $\mathbb{C}^n$ is due to
M. Marcus and B. N. Shure (1979). "The Numerical Range of Certain 0, 1-Matrices," *Linear and Multilinear Algebra* **7**, 111–120.

The calculation of the eigenvalues of the matrix B in the same example can be found in
M. Marcus (1960). "Basic Theorems in Matrix Theory," *Nat. Bur. Standards Appl. Math.*, Sec. 57.

Numerical ranges of weighted shifts were calculated by
Q. F. Stout (1983). "The Numerical Range of a Weighted Shift," *Proc. Amer. Math. Soc.* **88**, 495–502.
W. C. Ridge (1976). "Numerical Range of a Weighted Shift with Periodic Weights," *Proc. Amer. Math. Soc.* **55**, 107–110.

The calculation of the numerical radius of a 2×2 matrix, though straightforward, is quite involved. The pertinent formulas were summarized for the first time in
C. R. Johnson, I. M. Spitkovsky and S. Gottlieb (1994). "Inequalities Involving the Numerical Radius," *Linear and Multilinear Algebra* **37**, 13–24.

A general form of Theorem 1.3-4 appeared in
R. Bouldin (1971). "The Numerical Range of a Product II," *J. Math. Anal. Appl.* **33**, 212–219.

The reducing subspaces of Theorem 1.3-5 for operators with $2w(T) = \|T\|$ were first given by
J. P. Williams and T. Crimmins (1974). "On the Numerical Radius of a Linear Operator," *Amer. Math. Monthly* **74**, 832–833.

The best constant C for the inequality $\|T^n f\| \leq C\|f\|$ of Theorem 1.3-6 was found to be $\sqrt{2}$ by
M. J. Crabb (1971). "The Powers of an Operator of Numerical Radius One," *Mich. Math. J.* **18**, 252–256.

Earlier work on this problem appears in
C. A. Berger and J. G. Stampfli (1967). "Norm Relations and Skew Dilations," *Acta. Sci. Math.* (Szeged) **28**, 191–195.
T. Kato (1965). "Some Mapping Theorems for the Numerical Range," *Proc. Japan Acad.* **41**, 652–655.

The finite dimensional version of this inequality is rendered more easily using the theorem on the ascent of an operator proved by
N. Nirschl and H. Schneider (1964). "The Bauer Field of Values of a Matrix," *Numer. Math.* **6**, 355–365.

For more details, see [BD].

Theorem 1.3-8 and further consequences of $w(T) \leq 1$ are due to
T. Furuta and R. Nakamoto (1971). "Certain Numerical Radius Contraction Operators," *Proc. Amer. Math. Soc.* **29**, 521–524.

The numerical radius of nilpotent operators is discussed in

U. Haagerup and P. de la Harpe (1992). "The Numerical Radius of a Nilpotent Operator on a Hilbert Space," *Proc. Amer. Math. Soc.*. **115**, 371–379.

1.4 Normal Operators

From Theorems 1.2-3 and 1.2-4, we see that for a selfadjoint operator T, we have

$$r(T) = w(T) = \|T\|,$$

i.e., equality of spectral, numerical and operator radii. This property generalizes to normal operators, as will be seen in the following. In turn, the properties of a normal operator related to its numerical range have served as the source of a variety of generalizations to other operator classes, which will be developed further in Chapter 6. Recall that *normal operators*, those T for which $T^*T = TT^*$, may be regarded as a generalization of selfadjoint operators T in which T^* need not be exactly T but commutes with T.

Theorem 1.4-1. *If $W(T)$ is a line segment, then T is normal.*

Proof. Let α be a point on the line segment with inclination θ. Then $W(e^{-i\theta}[T-\alpha I])$ is contained in the real axis. Thus $e^{-i\theta}[T-\alpha I]$ is selfadjoint and so T is normal. □

Theorem 1.4-2. *If T is normal, then $\|T^n\| = \|T\|^n$, $n = 1, 2, \ldots$. Moreover, then*

(1.4-1) $$r(T) = w(T) = \|T\|.$$

Proof. For any $x \in H$,

(1.4-2) $$\|Tx\|^2 = \langle T^*Tx, x \rangle \leq \|T^*Tx\|.$$

Hence $\|T\|^2 \leq \|T^2\|$. Since we always have $\|T^2\| \leq \|T\|^2$, we conclude that $\|T^2\| = \|T\|^2$. Now, $\|T^n x\|^2 = \langle T^*T^n x, T^{n-1}x \rangle \leq \|T^{n+1}x\| \|T^{n-1}x\|$ for $n = 2, 3, \ldots$. Combining this result with $\|T\|^2 = \|T^2\|$ and using induction, we prove that $\|T^n\| = \|T\|^n$, $n = 1, 2, \ldots$. Moreover, recalling that $r(T) = \lim_{n \to \infty} \|T^n\|^{1/n}$, since $\|T^n\| = \|T\|^n$, we have $r(T) = \|T\|$ and (1.4-1). □

Theorem 1.4-3. *Let z be any complex number in the resolvent set of a normal operator T. Then*

(1.4-3) $$\|(T - zI)x\| \geq d(z, \sigma(T)) \quad \text{for} \quad x \in H, \|x\| = 1.$$

Proof. Since $(T - zI)^{-1}$ is normal, we have
$$\|(T - zI)^{-1}\| = r((T - zI)^{-1}),$$
and so
$$\|(T - zI)^{-1}\| = \sup\{|\lambda|, \ \lambda \in \sigma\{(T - zI)^{-1}\}\}.$$
Using the spectral mapping theorem, we obtain

(1.4-4) $$\|(T - zI)^{-1}\| = \frac{1}{d(z, \sigma(T))}.$$

Thus for any $x \in H$, $\|x\| = 1$, we have

(1.4-5) $$d(z, \sigma(T)) = \|(T - zI)^{-1}\|^{-1} \leq \|(T - zI)x\|. \quad \square$$

Theorem 1.4-4. *The closure of the numerical range of a normal operator is the convex hull of its spectrum.*

Proof. We need only to prove that any closed half-plane in $\mathbb{C}$ containing $\sigma(T)$ also contains $W(T)$. Without loss of generality, we can assume that $\sigma(T) \subset \{\lambda : \operatorname{Re} \lambda \leq 0\}$ and that the imaginary axis is a supporting line for $\operatorname{co}(\sigma(T))$.

Suppose that $a + ib \in W(T)$ with $a > 0$ and $\langle Tx, x \rangle = a + ib$, $\|x\| = 1$. Let $Tx = (a + ib)x + y$, where $\langle x, y \rangle = 0$. Let $c \in \mathbb{R}$, $c > 0$. Then $c \notin \sigma(T)$ and we have
$$d(c, \sigma(T)) \leq \|(T - cI)x\|,$$
i.e.,
$$c^2 \leq \|(a - c + ib)x + y\|^2 = (a - c)^2 + b^2 + \|y\|^2.$$
Hence $2ac \leq a^2 + b^2$ with $a, c > 0$. This is impossible since c is arbitrary. $\quad \square$

The following theorem provides a necessary and sufficient condition for the closedness of $W(T)$ of a normal operator in terms of the extreme points of $W(T)$. A point z is an *extreme point* of a set S if $z \in S$ and there is a closed half-plane containing z and no other element of S.

Theorem 1.4-5. *The extreme points of the closure of the numerical range $W(T)$ of a normal operator T are eigenvalues of T if and only if $W(T)$ is closed.*

Proof. Let $W(T)$ be closed. We can assume that the extreme point is $z = 0$ and that $W(T) \subseteq \{\lambda : \operatorname{Im} \lambda \geq 0\}$ and $\langle Tx, x \rangle = 0 \in W(T)$; hence $\langle (T - T^*)x, x \rangle = 0$. Since the operator $\frac{1}{i}(T - T^*) \geq 0$, it follows that $(T - T^*)x = 0$. Consequently, x is an element of the closed subspace $\{f : Tf = T^*f\} = N$. Since T is normal, we have
$$T^*Tx = TT^*x = TTx,$$

and hence the subspace N is invariant for T and $T|_N$ is selfadjoint. Obviously, $W(T|_N) \subset W(T)$ and $W(T|_N) \subset \mathbb{R}$ by Theorem 1.2-2. Hence $W(T|_N) \subset W(T) \cap \mathbb{R} = \{0\}$, and thus $T|_N = 0$ and $Tx = 0$, i.e., $0 \in \sigma_p(T)$.

The converse is true for any operator T. The compact convex set $\overline{W(T)}$ is the convex hull of its extreme points. When the latter are eigenvalues of T, as assumed in the theorem, we have

$$\overline{W(T)} \subset \mathrm{co}(\sigma_p(T)) \subset \mathrm{co}(W(T)) = W(T). \quad \square$$

Notes and References for Section 1.4

The bound of Theorem 1.4-3 was observed by
G. H. Orland (1964). "On a Class of Operators," *Proc. Amer. Math. Soc.* **15**, 75–79.

The equality of Theorem 1.4-4 of the convex hull of the spectrum and the numerical range was first mentioned in
M. H. Stone (1932). *Linear Transformations in Hilbert Space*, American Mathematical Society, R.I.
This property was established later in
S. K. Berberian (1964). "The Numerical Range of a Normal Operator," *Duke Math. J.* **31**, 479–483.

The proof we adopted in Theorem 1.4-5 is due to
J. G. Stampfli (1966). "Extreme Points of the Numerical Range of a Hyponormal Operator," *Michigan Math. J.* **13**, 87–89.
An earlier proof appears in
C. H. Meng (1957). "A Condition That a Normal Operator Has a Closed Numerical Range," *Proc. Amer. Math. Soc.* **8**, 85–88.

The relationship of Theorem 1.4-5 to Theorem 1.3-3 should be noted. The point is that the numerical ranges of finite-dimensional normal operators are polygons whose vertices are eigenvalues.

Resolvent estimates are key to many aspects and applications of operator theory. In particular, certain operator classes have been defined in terms of them. These classes will be discussed in Chapter 6. Also, they play an important role in numerical analysis, as will be seen in Chapter 4. For normal operators, such estimates sometimes become exact, as in (1.4-4), for example. For the general case, the following should therefore be kept in mind:

$$(1.4\text{-}6) \qquad d(\lambda, \sigma(T)) \geqq \|(T - \lambda I)^{-1}\|^{-1} = \inf_x \frac{\|(T - \lambda I)x\|}{\|x\|}.$$

1.5 Numerical Boundary

A natural question regarding the boundary of the numerical range is: which points of $W(T)$ are on the boundary? In this section we will characterize them and further obtain some properties of special extreme points.

For each complex number z, let M_z be the subset of H given by

(1.5-1) $$M_z = \{x : \langle Tx, x \rangle = z\|x\|^2\}.$$

Note that M_z is homogeneous, closed and not necessarily linear. Let sp M_z denote the linear span of M_z.

Theorem 1.5-1. *If $z \in W(T)$ is an extreme point of $W(T)$, then M_z is linear.*

Proof. Let L be the line of support of $W(T)$ at z. Then, for some θ, $\operatorname{Re} \langle (e^{-i\theta}(T - zI)x, x \rangle \geqq 0$ and

(1.5-2) $$M_z = \{x : \operatorname{Re} \langle (T - zI)x, x \rangle = 0\}.$$

For $x, y \in M_z$,

$$\langle e^{-i\theta}(T - zI)(x + y), (x + y) \rangle = a = -\langle e^{-i\theta}(T - zI)(-x + y), (-x + y) \rangle$$

is purely imaginary. If $a \neq 0$, we have purely imaginary elements of $W(e^{-i\theta}(T - zI))$ in both the upper and lower half-planes, and z is not an extreme point. So $a = 0$ and $x + y \in M_z$. □

Theorem 1.5-2. *$z \in W(T)$ is an extreme point if M_z is linear.*

Proof. Suppose that z is not an extreme point. Thus z is an interior point of the line segment joining $a, b \in W(T)$ with $\langle Tx, x \rangle = a$ and $\langle Ty, y \rangle = b$, $\|x\| = \|y\| = 1$. It can be shown that there exist $t \in (0, 1)$ and a complex λ, $|\lambda| = 1$, such that $tx + (1 - t)\lambda y \in M_z$. Since $\langle T(-\lambda y), (-\lambda y) \rangle = b$, we can similarly assure that for the same λ there is an $s \in (0, 1)$ such that $sx + (1 - s)(-\lambda y) \in M$, so $x \in M_z$. But $a \neq z$ and $M_a \cap M_z = \{0\}$, a contradiction. Hence z is an extreme point. □

Theorem 1.5-3. *Let z be a nonextreme boundary point of $W(T)$ and L the line of support at z with inclination θ. Then*

$$N = \bigcup \{M_\alpha : \alpha \in L\}$$

is a closed subspace of H and sp $M_z = N$.

Proof. For all $\alpha \in L$, $\arg(\alpha - z) = \theta$ is constant and $e^{-i\theta}(\alpha - z)$ is real. Consequently,

$$N = \{x : \langle e^{i\theta}(T - zI)x, x \rangle \text{ is real}\}.$$

1.5 Numerical Boundary

Since L is a line of support, we have either $\operatorname{Im} e^{-i\theta}(T-zI) \geq 0$ or $\operatorname{Im} e^{i\theta}(T-zI) \leq 0$. Assuming the former, we have a selfadjoint operator $\operatorname{Im} e^{i\theta}(T-zI) = B$ and

$$N = \{x : \langle Bx, x\rangle = 0\} = \{x : Bx = 0\}$$

or equivalently,

$$N = \{x : e^{i\theta}(T-zI)x = e^{i\theta}(T^*-zI)x\}.$$

So N is a closed subspace of H. As in Theorem 1.5-2, we have for each $\alpha \in L$,

(1.5-3) $$M_\alpha \subset M_z + M_z = \operatorname{sp} M_z,$$

and hence $N \subset \operatorname{sp} M_z$. Since $M_z \subset N$ and N is a subspace, we have $\operatorname{sp} M_z \subset N$. □

So far, we have looked at $\partial W(T) \cap W(T)$. Attempts to describe the unattained boundary points $\overline{W(T)} - W(T)$ lead to the consideration of special sequences of vectors in H. Notice that for any $z \in \partial W(T)$ there is always a sequence of unit vectors $\{f_n\}$ such that $\{\langle Tf_n, f_n\rangle\} \to z$. Since the unit sphere is weakly compact, by taking subsequences, we can assume that $f_n \xrightarrow{w} f$ and $\langle Tf_n, f_n\rangle \to z$. Let us now define a special property (P) of such sequences as follows. Let $\{f_n\}$ be a sequence of unit vectors in H, $f_n \xrightarrow{w} f$ and $\langle Tf_n, f_n\rangle \to z$. Then $\{f_n\}$ is said to have *property (P)* if either $f = 0$ or $\langle Tf, f\rangle = z\|f\|^2$.

Theorem 1.5-4. *If z is an extreme point of $\overline{W(T)}$ and $\langle Tf_n, f_n\rangle \to z$, $\|f_n\| = 1$, then the sequence $\{f_n\}$ has the property (P).*

Proof. Let $g_n = f_n - f$, $g_n \xrightarrow{w} 0$, and $\|g_n\| \leq 2$. Hence, by considering subsequences, we may assume that $\|g_n\| \to a \in \mathbb{R}$. Since $1 = \|f_n\|^2 = \|g_n + f\|^2 = \|g_n\|^2 + \|f\|^2 + 2\operatorname{Re}\langle g_n, f\rangle$, we have

$$\langle Tf_n, f_n\rangle = \langle Tg_n, g_n\rangle + \langle g_n, T^*f\rangle + \langle Tf, g_n\rangle + \langle Tf, f\rangle \to z.$$

Since $\{g_n\} \xrightarrow{w} 0$, $\langle Tg_n, g_n\rangle \to z - \langle Tf, f\rangle$.

If $a \neq 0$, let $\langle Tf, f\rangle = \alpha\|f\|^2$ and

$$\frac{\langle Tg_n, g_n\rangle}{\|g_n\|^2} = \beta_n, \qquad g_n \neq 0.$$

Then $\beta_n \to \lambda \in \overline{W(T)}$, and we have $\langle Tg_n, g_n\rangle = \lambda a^2$ and $z = \alpha\|f\|^2 + \lambda a^2$. Since $a^2 + \|f\|^2 = 1$, we have that z lies on the segment joining α and λ, both belonging to $\overline{W(T)}$. So z cannot be an extreme point. Thus, either $\lambda = z$ or $\alpha = z$. In either case, $z = \alpha$ and $\langle Tf, f\rangle = z\|f\|^2$. So $\{f_n\}$ has the property (P). □

A point $z \in \overline{W(T)}$ is called a *corner* if $\overline{W(T)}$ is contained in a half-cone with vertex at z, and the semivertical angle of the cone is less than $\frac{\pi}{2}$.

Theorem 1.5-5. *If $z \in W(T)$ is a corner of $\overline{W(T)}$, then $z \in \sigma_p(T)$.*

Proof. Let z be a corner and $\langle Tf, f \rangle = z$, $\|f\| = 1$. If $Tf \neq \lambda f$ for any λ, then f and Tf are linearly independent. Let E be the orthogonal projection onto the subspace spanned by f and Tf. $W(ETE)$ is an ellipse contained in $W(T)$. Since $z \in W(ETE)$, z is contained in the ellipse. This is impossible since z is a corner unless the ellipse is degenerate and z is an eigenvalue μ of ETE. But then $ETEf = \mu f = Tf$, a contradiction. □

Corollary 1.5-6. *If $z \in \overline{W(T)}$ is a corner of $\overline{W(T)}$, then $z \in \sigma_{\text{app}}(T)$.*

Proof. Embed (see Notes) H in a larger Hilbert space K and extend T to a bounded linear operator $\bar{T}$ on K such that $\sigma_{\text{app}}(T) = \sigma_p(\bar{T})$. Then $\bar{T}x = \lambda x$ for some $x \neq 0$. If $x_n \to x$ and $Tx_n \to \bar{T}x$, we have $\|(T - \lambda I)x_n\| \to 0$. □

Corollary 1.5-7. *If $\overline{W(T)}$ is a closed polygon, $W(T) = \text{co}\,\sigma(T)$.*

Proof. By Theorem 1.5-5, the vertices of the polygon are in $\sigma(T)$. □

Corollary 1.5-8. *A compact polygon with m vertices is the numerical range of an operator in an n-dimensional space iff $m \leq n$.*

Proof. If the numerical range polygon has m vertices, they are corners. By Theorem 1.5-5, each of them is in $\sigma_p(T)$ and there can be at most n of them, by the linear independence of their corresponding eigenvectors. Conversely, if we consider the normal operator T represented by the matrix $A = [a_{ij}]$ with elements

$$\begin{cases} a_{ij} = \lambda_i \delta_{ij} & (1 \leq i \leq m), \\ a_{ij} = 0 & m < i \leq n, \end{cases}$$

then $W(T) = \text{co}\,\sigma(T)$. □

Notes and References for Section 1.5

The geometry of the numerical range and its boundary, especially $W(T) - \text{Int}\,W(T)$, was studied by
J. Agler (1982). "Geometric and Topological Properties of the Numerical Range," *Indiana Univ. Math. J.* **31**, 766–767.
 The fact that any subset G of $\mathbb{C}$ can be the numerical range of some bounded operator, if $G - \text{Int}\,G = E_0 \cup E_1$ where E_0 is countable and E_1 is a union of smooth subarcs of a conic section, was shown by
M. Radjabalipour and H. Radjavi (1975). "On the Geometry of Numerical Range," *Pacific J. Math.* **61**, 507–511.

1.5 Numerical Boundary

The definition of M_z and their use in relating extreme points are studied by

M. Embry (1970). "The Numerical Range of an Operator," *Pacific J. Math.* **32**, 647–650.

Notice that both of the Theorems 1.5-1 and 1.5-2 need $z \in W(T)$. The case when $z \in \overline{W(T)} - W(T)$ was studied by

B. Sims (1972). "On the Connection Between the Numerical Range and Spectrum of an Operator in a Hilbert Space," *J. London. Math. Soc.* **8**, 57–59.

K. C. Das (1977). "Boundary of Numerical Range," *J. Math. Anal. Appl.* **60**, 779–780.

and

G. Garske (1979). "The Boundary of the Numerical Range of an Operator," *J. Math. Anal. Appl.* **68**, 605–607.

We will speak more about dilation theory in the next chapter. However, the embedding argument of Corollary 1.5-6 follows from a construction in

S. K. Berberian (1962). "Approximate Proper Vectors," *Proc. Amer. Math. Soc.* **113**, 111–114.

S. K. Berberian and G. H. Orland (1967). "On the Closure of the Numerical Range of an Operator," *Proc. Amer. Math. Soc.* **18**, 499–503.

The case of compact operators was studied extensively by

G. de Barra, J. R. Giles and B. Sims (1972). "On the Numerical Range of Compact Operators on Hilbert Spaces," *J. Lond. Math. Soc.* **5**, 704–706.

That the extreme points of $W(T)$ lie in the spectrum for a normal operator T was shown by

C. R. Macluer (1965). "On Extreme Points of the Numerical Range of Normal Operators," *Proc. Amer. Math. Soc.* **16**, 1183–1184.

The corresponding result for a hyponormal operator (see Chapter VI) was given by

J. G. Stampfli (1966). "Extreme Points of the Numerical Range of Hyponormal Operators," *Michigan Math. J.* **13**, 87–89.

The same result for compact operators was given in

G. de Barra (1981). "The Boundary of the Numerical Range," *Glasgow Math. J.* **22**, 69–72.

Much earlier

W. F. Donoghue (1957). "On the Numerical Range of a Bounded Operator," *Michigan Math. J.* **4**, 261–263.

proved that when $W(T)$ is closed, a point on the boundary at which it is not a differentiable arc is an eigenvalue of T.

1.6 Other W-Ranges

The concept of the numerical range $W(T)$ originated from the real line segment $[\lambda_{\min}, \lambda_{\max}]$ of quadratic form values $\langle Tx, x \rangle$ of a symmetric matrix T, the continuous segment being fully obtained by the quadratic form values as x ranges over the unit sphere $\|x\|^2 = 1$. The numerical range $W(T)$ defined the same way for arbitrary bounded linear operators T on a finite- or infinite-dimensional Hilbert space is now an established chapter in Hilbert space linear operator theory.

Generalizations from there have gone in roughly two directions. From the matrix analysis viewpoint, a number of variations on $W(T)$ have been investigated in finite dimensions. From the operator theory viewpoint, the principal generalization has been the extension to Banach spaces based upon the Hahn–Banach theorem and the notion of a semi-inner product.

The matrix analysis variations that have been and are currently being studied include the k-numerical range (a joint numerical range); its generalization, the C-numerical range (which has recently received the most attention); what we may call the sesquilinear-numerical range $\{\langle A^{1/2}Tx, A^{1/2}x \rangle, \|x\|^2 = 1\}$, which is just the usual numerical range of T but now in the inner product coming from the sesquilinear form $\langle Ax, x \rangle$, where A is any specified positive definite operator (this is sometimes called the generalized field of values); the M numerical range (based on resolvent growths); restricted numerical ranges (in which x is restricted to only portions of the unit sphere in the Hilbert space H); symmetrizations of the numerical range; and others. These will be discussed further in Chapter 5.

The Banach space generalization in operator theory relies on the *Hahn–Banach theorem*, from which a standard interesting corollary is as follows.

Lemma 1.6-1. *For every vector x in a normed linear space X, there exists a linear functional x^* in the dual space X^* such that*

$$x^*x = \|x\|^2 = \|x^*\|^2.$$

From this we may define a semi-inner product. A *semi-inner product* is a mapping $[\,,\,]$ from $X \times X$ to the scalars such that
  (i) $[x, y]$ is linear in x for each fixed y,
  (ii) $[x, x] > 0$ when $x \neq 0$,
  (iii) $|[x, y]|^2 \leq [x, x][y, y]$,
We shall also always assume that the semi-inner product is consistent with the norm: $[x, x] = \|x\|^2$. Clearly, properties (i)–(iii) generalize those of the usual Hilbert space inner product, the only loss being the bilinearity.

Example. Given a Banach space X, let a semi-inner product $[x, y]$ be defined by

$$[x, y] = y^*x,$$

where y^* is a linear functional corresponding to y as guaranteed in Lemma 1.6-1.

Whether or not a Banach space X has more than one such semi-inner product depends on the geometry of the unit balls in X and X^*. For any chosen semi-inner product, one then defines the *spatial numerical range* as usual,

$$W(T) = \{[Tx, x];\ \|x\|^2 = 1\}.$$

When there is more than one semi-inner product, one defines the *(total) spatial numerical range* to be

$$(1.6\text{-}1) \qquad V(T) = \bigcup_\alpha W_\alpha(T),$$

where α indexes all semi-inner products on X. These Banach space numerical ranges (and similar Banach algebra variations) are fully exposited in [BD]. In matrix theory, $V(A)$ is called $F(A)$, the Bauer field of values.

Important properties for everywhere defined, bounded operators T include the following: the closure of $V(T)$ contains not only the spectrum $\sigma(T)$ but also its convex hull; $V(T)$ need not be convex but is connected; and the numerical radius $|V(T)|$ is (complex scalar field assumed now, of course) an equivalent norm and obeys the inequalities

$$(1.6\text{-}2) \qquad |V(T)| \leq \|T\| \leq e|V(T)|.$$

The spectral inclusion property fully generalizes that of the Hilbert space case, Theorem 1.2-1. The connectedness property captures the essence of Theorem 1.1-2. The norm equivalence inequality, (1.6-2), corresponds to that of Theorem 1.3-1, with the ratio 2 replaced by e, which is known to be sharp for Banach spaces. Many of the normal and numerical range boundary properties of Sections 1.4 and 1.5, respectively, have also been generalized to some extent to operators T on Banach spaces and in Banach algebras. See [BD] for more information.

Notes and References for Section 1.6

Generalizations of the numerical range in the finite-dimensional case are summarized in [HJ2]. We shall discuss their properties further in Chapter 5. An early influential paper was that of
F. Bauer (1962). "On the Field of Values Subordinate to a Norm," *Numerische Math.* **4**, 103–111.

Because the Banach space generalizations are treated in full in [BD], we shall not treat them in this book. Let us note that the [BD] expositions are further updated in
F. Bonsall and J. Duncan (1980). "Numerical Ranges," in *Studies in Mathematics* **21**, ed. R. G. Bartle, Mathematical Association of America, 1–49.

24   1. Numerical Range

For a speculation that the $|V(T)|$ radius and operator norm $\|T\|$ must go to infinity together, i.e., that T is bounded if and only if the numerical range $V(T)$ is bounded, see

K. Gustafson and B. Zwahlen (1974). "On Operator Radii," *Acta Sci. Math.* **36**, 63–68.

This proposition is probably related to still not fully understood connectivity properties of $V(T)$.

For relationships between the semi-inner product structures of Banach spaces and partial inner product structures of rigged Hilbert spaces, and a far reaching generalization of such structures to Galois connections, see

J. P. Antoine and K. Gustafson (1981). "Partial Inner Product Spaces and Semi-inner Product Spaces," *Advances in Mathematics* **41**, 281–300.

Endnotes for Chapter 1

Although our general stance in this book will be that of bounded, everywhere defined linear operators on a complex Hilbert space, it is worth keeping in mind three different general contexts for the numerical ranges $W(T)$:

  1. finite-dimensional case, T given by a matrix. $W(T)$ is a compact (closed and bounded) convex set;

  2. infinite-dimensional case, T bounded and given, for example, by an infinite matrix or an integral. $W(T)$ is convex, bounded, but not necessarily closed;

  3. infinite-dimensional case, T unbounded (either closed or unclosed) and given, for example, by a singular integral or a differential equation. $W(T)$ is convex but unbounded and not necessarily closed.

In all three contexts (taking T closed and densely defined in the third, to keep it simple), the spectrum $\sigma(T)$ is a closed set, a great convenience indeed. However, the spectrum $\sigma(T)$ need not be contained in $W(T)$ in the third instance, and the spectrum $\sigma(T)$ need not be a point spectrum (i.e., eigenvalues) in the second instance. There are other distinctions between these three situations, and we don't wish to belabor them. Our point is that the "operator theory community" is really three communities. Their outlooks differ. Most importantly, the methods of proof may also differ, due to distinctions in the three situations such as those we have just noted.

Often, nearly identical theorems have been proved independently by the three communities. It is beyond our effort here to be sure that we have identified those instances. In fact we believe that there are sufficiently many such instances that we decided not to even try to account for them in any systematic way. However, let us give one important example of this so that our point is clear. The assumption that

$$0 \notin \overline{W(T)}$$

is strong. It means that a complex rotation $e^{i\theta}$ of T has numerical range (and spectrum) contained in a closed right half-plane $\operatorname{Re} z \geqq \lambda_0 > 0$. This assumption has yielded a number of interesting results in matrix theory and continues to do so. From the other viewpoint of operator theory, such operators, generally unbounded, are called (when rotated) *m-accretive* operators and have been treated extensively as infinitesimal generators in the theory of operator semigroups. The weaker condition that $\operatorname{Re}\langle Tx, x\rangle \geqq 0$, already used in the proof of Theorem 1.5-1 and to be used extensively in the next chapter and often in later chapters, is called *accretive* in the semigroup theory, where it is equally useful.

One of the beauties of the numerical range $W(T)$, just as with the spectrum $\sigma(T)$, is that its use and properties do indeed extend over all three contexts: matrix analysis, operator theory, and differential equations. Our goal in writing this book is to expose the basic properties of $W(T)$ common to all three contexts, principally within the setting of bounded operators T on a complex Hilbert space H.

*　　　　　*　　　　　*

As described in Section 1.6, there is a fourth context for the numerical range: the generalized numerical range $V(T)$ for T a densely defined operator in a Banach space. Actually, that context itself divides into two cases:

4. Banach space case: finite- or infinite-dimensional T, bounded or unbounded, spatial numerical range $V(T)$;

5. Banach algebra case: T an element of a normed unital algebra, algebraic numerical range $v(T)$.

Of course, there are strong overlaps between contexts 4 and 5, but the outlooks are significantly different. In the first, T is usually regarded as known to us, and one explores the various properties of $V(T)$, especially as they may resemble or differ from those of $W(T)$ in the Hilbert space case. In the second, the normed algebras are regarded as the items of interest, and the algebraic numerical range is regarded as a tool with which to learn more about the algebra.

There is in fact a sixth context for the numerical range:

6. Nonlinear operator theory: T a nonlinear operator in a Hilbert or Banach space.

Some of the $W(T)$ and $V(T)$ results extend to nonlinear operators T in interesting ways. For example, the Bonsall, Cain and Schneider (see [BD]) result that $V(T)$ remains connected depends only on the continuity of T and not on its linearity. However, we have taken the view that the core theory is linear, and indeed nonlinear operator theories are often quite specialized. Examples of the nonlinear numerical range theories may be found in

E. Zarantonello (1967). "The Closure of the Numerical Range Contains the Spectrum," *Pacific J. Math.* **22**, 575–595.

H. Brezis (1973). *Operateurs Maximaux Monotones et Semi-groups de Contractions dans les Espaces de Hilbert*, North Holland, Amsterdam.

V. Barbu (1976). *Nonlinear Semigroups and Differential Equations in Banach Spaces*, Editura Academie, Bucharest, Rumania.

I. Miyadera (1992). *Nonlinear Semigroups*, Amer. Math. Soc., Providence, R.I.

2
Mapping Theorems

Introduction

Mapping theorems for the numerical range analogous to the spectral mapping theorem are hard to come by. The analogy is rather limited by the convexity of the numerical range. However, significant results were obtained in relating the numerical ranges and the numerical radii of an operator T to those of the operator $f(T)$, where f is is a given function. As can be expected, the best results were obtained in the special case $f(T) = T^n$, n a natural number. In addition to the preceding results, this chapter also gives some other results for the numerical range of products, commuting operators, and the natural connections between the numerical range and the theory of dilations of operators.

Let us first observe that even $W(T)$ and $W(T^2)$ need not be related in a simple way. We can see in the following example that $W(T)$ is contained in the sector $\{-\frac{\pi}{4} \leq \theta \leq \frac{\pi}{4}\}$ and $W(T^2)$ is not contained in the sector $\{-\frac{\pi}{2} \leq \theta \leq \frac{\pi}{2}\}$.

Example. Let $H = \mathbb{C}^2$ and

$$T = \begin{bmatrix} 4+i & 4i \\ 4i & 16+4i \end{bmatrix}.$$

For $u = (x, y) \in \mathbb{C}^2$, we have

$$\operatorname{Re} \langle Tu, u \rangle = 4|x|^2 + 16|y|^2,$$

$$\operatorname{Im} \langle Tu, u \rangle = 3|x|^2 + \frac{9}{2}|y|^2 < \operatorname{Re} \langle Tu, u \rangle.$$

However, for $u = (1, 0)$, $\langle T^2 u, u \rangle = -1 + 8i$.

2.1 Radius Mapping

Simple relations between $w(T)$ and $w(f(T))$ can be obtained in special cases. The best-known example is the power inequality.

Theorem 2.1-1 (Power inequality). *Let T be an operator and $w(T) \leq 1$. Then $w(T^n) \leq 1$, $n = 1, 2, 3, \ldots$.*

Proof. Let us first observe that for any $z \in \mathbb{C}$, $|z| < 1$,
$$\operatorname{Re} \langle (I - zT)x, x \rangle = \|x\|^2 - \operatorname{Re} \langle zTx, x \rangle \geq \|x\|^2 [1 - |z|] \geq 0$$
follows from $w(T) \leq 1$. On the other hand, when $\operatorname{Re}\langle (I - zT)x, x \rangle \geq 0$ for all $z \in \mathbb{C}$, $|z| < 1$, we have, taking $z = te^{i\alpha}$ and letting $t \to 1$, $\operatorname{Re} \langle e^{i\alpha}Tx, x \rangle \leq \|x\|^2$ and hence $w(T) \leq 1$. Thus, whenever $I - zT$ is invertible, we have $\operatorname{Re} \langle (I - zT)x, x \rangle \geq 0$, $\forall x \in H \Leftrightarrow \operatorname{Re} \langle (I - zT)^{-1}y, y \rangle \geq 0$, $\forall y \in H$, using $x = (I - zT)^{-1}y$. Notice that $r(T) \leq 1$ and that $I - zT$ is invertible.

Thus, it is sufficient to prove that for all $x \in H$
$$\operatorname{Re} \langle (I - z^n T^n)^{-1} x, x \rangle \geq 0 \quad \text{with} \quad z \in \mathbb{C}, |z| < 1.$$
We now use the identity
$$(I - z^n T^n)^{-1} = \frac{1}{n} \big[(I - zT)^{-1} + (I - \omega zT)^{-1} + \cdots + (I - \omega^{n-1} zT)^{-1} \big],$$
where ω is a primitive nth root of 1. Since for each of the operators $(I - \omega^k zT)^{-1}$ we have
$$\operatorname{Re} \langle (I - \omega^k zT)^{-1} x, x \rangle \geq 0 \quad \text{for all} \quad x \in H,$$
we deduce that
$$\operatorname{Re} \langle (I - z^n T^n)^{-1} x, x \rangle \geq 0 \quad \text{for all} \quad x \in H.$$
Thus $\operatorname{Re} \langle (I - z^n T^n)x, x \rangle \geq 0$, $\forall x \in H$, $|z| < 1$, and therefore $w(T^n) \leq 1$. □

Conditions equivalent to $w(T) \leq 1$, like the one in the preceding proof,

(2.1-1) $\quad w(T) \leq 1 \Longleftrightarrow \operatorname{Re} \langle (I - zT)x, x \rangle \geq 0 \Longleftrightarrow \operatorname{Re} \langle (I - zT)^{-1}x, x \rangle \geq 0,$

for $x \in H$, $|z| < 1$, play an important role in mapping theorems for the numerical radius. Some other obvious equivalent conditions are

(2.1-2) $\quad\quad\quad \operatorname{Re} \langle (I + zT)x, x \rangle \geq 0, \quad \forall x \in H, |z| < 1,$
(2.1-3) $\quad\quad\quad \operatorname{Re} \langle zT(I - zT)^{-1}x, x \rangle \geq 0, \quad \forall x \in H, |z| < 1.$

See the Notes and References for further equivalent conditions.

The following theorem provides a dilation condition equivalent to $w(T) \le 1$, which is quite useful in mapping theorems.

Theorem 2.1-2. (Power dilation) $w(T) \le 1$ if and only if $T^n = 2PU^n P$ for $n = 1, 2, \ldots$, where U is a unitary operator on a Hilbert space $K \supset H$ and P is the projection of K on H.

Proof. Assuming that $w(T) \le 1$, we have $r(T) \le 1$, and the operator $(I - zT)^{-1}$ exists for $|z| < 1$. The operator valued function $F(z) = (I - zT)^{-1}$ is holomorphic in the disk $\{|z| < 1\}$, $F(0) = I$ and $\operatorname{Re} F(z) \ge 0$. It follows (see the Notes and References; see also Section 2.6) by a theorem of Riesz, generalized to operator valued functions, that there exist a Hilbert space $K \supset H$ and a unitary operator U in K such that
$$F(z) = P(I_K + zU)(I_K - zU)^{-1}, \quad |z| < 1.$$
By series expansion,
$$\begin{aligned} F(z) &= P[I + 2zU + 2z^2 U^2 + \cdots] \\ &= I_H + zT + z^2 T^2 + \cdots. \end{aligned}$$
Equating coefficients, we obtain

(2.1-4) $\qquad T^n = 2PU^n P \quad \text{for} \quad n = 1, 2, 3, \ldots.$

The series $I_H + zT + z^2 T^2 + \cdots$ converges for $|z| < 1$ and equals $(I_H - zT)^{-1}$. On the other hand, the existence of the Hilbert space K and that of the unitary operator U such that $T^n = 2PU^n P$ imply that the series $I_K + 2zU + 2z^2 U^2 + \cdots$ converges for $|z| < 1$. Consequently, the series
$$I_H + zT + z^2 T^2 + \cdots$$
converges in norm for $|z| < 1$ and equals $(I - zT)^{-1}$.

Evaluating
$$\langle (I_K + zU)(I_K - zU)^{-1} y, y \rangle$$
and taking $y = (I_K - zU)x$, we obtain

(2.1-5) $\qquad \begin{aligned} \operatorname{Re} \langle (I_K + U)(I_K - zU)^{-1} y, y \rangle &= \operatorname{Re} \langle (I_K + zU)x, (I_K - zU)x \rangle \\ &= (1 - |z|^2)\|x\|^2 \ge 0 \end{aligned}$

for $|z| < 1$. Hence $\langle (I - zT)^{-1} x, x \rangle \ge 0$ for $|z| < 1$. Thus $w(T) \le 1$, using (2.1-1). $\square$

One of the immediate applications of the preceding theorem above is to relate $w(T)$ and $w(f(T))$ when f is analytic.

Theorem 2.1-3. Let f be analytic in $|z| < 1$ and continuous on the boundary $|z| = 1$, with $f(0) = 0$. If $|f(z)| \le 1$ for $|z| \le 1$ and $w(T) \le 1$, then $w(f(T)) \le 1$.

Proof. We first prove that $\lim_{r \to 1} f(rT)$ exists in the norm. Let $U = \int e^{i\lambda} dE(\lambda)$ and $T^n = 2PU^n P$ for $n = 1, 2, \ldots$. Then

(2.1-6)
$$f(rT) = \Sigma a_n r^n T^n = 2P(\Sigma a_n r^n U^n)P$$
$$= 2P \left[\int \Sigma a_n r^n e^{in\lambda} dE(\lambda) \right] P$$
$$= 2P \left[\int f(re^{i\lambda}) dE(\lambda) \right] P.$$

By the continuity of f, we may conclude that

(2.1-7)
$$\lim_{r \to 1} f(rT) = 2P \left[\int f(e^{i\lambda}) dE(\lambda) \right] P$$
$$= 2P f(U) P.$$

It is easy to see that

$$[f(T)]^n = 2P[(fU)^n]P,$$

where $f(U)$ is a contraction and has a unitary dilation. Thus $w(f(T)) \leq 1$ using Theorem 2.1-2. $\square$

Notes and References for Section 2.1

The elementary proof in Theorem 2.1-1 is due to
C. Pearcy (1966). "An Elementary Proof of the Power Inequality for the Numerical Radius," *Mich. Math. J.* **13**, 289–291.
The final form of the proof is taken from [H]. The dilation condition equivalent to $w(T) \leq 1$ of Theorem 2.1-2 was found by
C. A. Berger (1965). "A Strange Dilation Theorem," *Amer. Math. Soc. Notices* **12**, 590.
The version used by us is due to
B. Sz.-Nagy and C. Foias (1966). "On Certain Classes of power-bounded operators in Hilbert Space," *Acta. Sci. Mat.* **27**, 17–25.

Conditions equivalent to $w(T) \leq 1$ have been studied by many authors. The conditions

$$w(T) \leq 1 \iff \operatorname{Re} zT \left[I - (1+\alpha)\frac{zT}{2} \right]^{-1} + I \geq 0,$$

for $|z| < 1$ and $|\alpha| < 1 \iff \|T - zI\| \leq 1 + \{1 + |z|^2\}^{1/2}$, $z \in C$, and Theorem 2.1-3 are due to
C. A. Berger and J. G. Stampfli (1967). "Mapping Theorems for the Numerical Range," *Amer. J. Math.* **89**, 1047–1055.

A characterization of operators with $w(T) \leq 1$ was obtained in
T. Ando (1973). "Structure of Operators with Numerical Radius One," *Acta. Sci. Mat.* (Szeged) **34**, 11–15.

Such T may be represented as $T = (I + A)^{1/2} B (I - A)^{1/2}$ where A is selfadjoint and A and B are contractions. This is equivalent to the existence of a contraction C such that $T = (I - C^*C)^{1/2} C$.

If $w(T) \leq 1$ and T is idempotent, then T is selfadjoint. If $T^K = T$ for $K \geq 2$, then T^{K-1} is a projection. For such facts, see T. Furuta and R. Nakamoto (1971). "Certain Numerical Radius Contraction Operators," *Proc. Amer. Math. Soc.* **29**, 521–524.

Minimum growth rates of operators T^n with $w(T) \leq 1$ were studied by E. S. W. Shiu (1976). "Growth of the Numerical Ranges of Powers of Hilbert Space Operators," *Mich. Math. J.* **23**, 155–160.

2.2 Analytic Functions

We will now look at some mapping theorems for the numerical range $W(T)$ and relate $W(f(T))$ and $W(T)$ when the function f is analytic in various regions.

Theorem 2.2-1. Let $f(z)$ be holomorphic on $\{z : |z| \leq 1\} = D$, and map D into $\{z : \operatorname{Re} z \geq 0\} = P$. If $W(A) \subset D$, then $W(f(A)) \subset P - \operatorname{Re} f(0)$.

Proof. The basic tool in the proof is the following expression for $f(z)$:

$$(2.2\text{-}1) \qquad f(z) = i \operatorname{Im} f(0) + \frac{1}{2\pi} \int_0^{2\pi} [\operatorname{Re} f(e^{it})] \frac{e^{it} + z}{e^{it} - z} \, dt,$$

which is a consequence of Herglotz's theorem (see Notes and References). Writing

$$(2.2\text{-}2) \qquad \frac{e^{it} + z}{e^{it} - z} = \frac{2e^{it} + z - e^{it}}{e^{it} - z} = \frac{2e^{it}}{e^{it} - z} - 1$$
$$= 2(1 - e^{-it} z)^{-1} - 1,$$

we have

$$(2.2\text{-}3) \qquad \begin{aligned} f(z) &= i \operatorname{Im} f(0) + \frac{1}{2\pi} \int_0^{2\pi} [\operatorname{Re} f(e^{it})][2(1 - e^{-it} z)^{-1} - 1] dt \\ &= \frac{1}{\pi} \int_0^{2\pi} [\operatorname{Re} f(e^{it})](1 - e^{-it} z)^{-1} dt - f(0). \end{aligned}$$

Replacing z by A and taking the real part, we have

$$(2.2\text{-}4) \qquad \operatorname{Re} f(A) = -\operatorname{Re} f(0) + \frac{1}{\pi} \int_0^{2\pi} [\operatorname{Re} f(e^{it}([I - e^{-it} A)^{-1}] dt,$$

Using (2.1-1), we have $\operatorname{Re}(I - e^{-it} A)^{-1} \geq 0$. As f maps D into P, we have $\operatorname{Re} f(e^{it}) \geq 0$. Hence, the integral on the right is positive and so

Re $f(A) \geq -\operatorname{Re} f(0)$. Thus
$$W(f(A)) \subset P - \operatorname{Re} f(0). \qquad \square$$

Corollary 2.2-2. *Let $f(z)$ be holomorphic on D, and map D into D. If $f(0) = 0$ and $W(A) \subset D$, then $W(f(A)) \subset D$.*

Proof. We can easily verify that the function $g(z) = \frac{1+af(z)}{1-af(z)}$, for any $a \in \mathbb{C}$, $|a| < 1$, maps D into P. Further, $g(0) = 1$. Hence, by Theorem 2.2-1, we have $W(g(A)) \subset P-1$. So Re $[g(A)+1] \geq 0$. However, $g(A)+1 = 2[I - af(A)]^{-1}$, and therefore Re $[I - af(A)] \geq 0$. Since a can be chosen arbitrarily, we conclude that
$$\operatorname{Re}[I - e^{i\theta}f(A)] \geq 0 \quad \text{for all} \quad \theta \in [0, 2\pi).$$
Thus $W(f(A)) \subset D$. $\qquad \square$

Notes and References for Section 2.2

The expression for $f(z)$ in (2.2-1) can be seen in
M. H. Stone (1932). *Linear Transformations in Hilbert Space*, American Mathematical Society, R.I.
   The Theorem 2.2-1 and Corollary 2.2-2 are given in
T. Kato (1965). "Some Mapping Theorems for the Numerical Range," *Proc. Japan Acad.* **41**, 652–655.
Recently, interesting results for the mapping of $f(A) = A^K$ were obtained by
C. K. Li (1995). "Numerical Ranges of the Powers of an Operator," Preprint.
The following is one of the important results:
   For any real $\phi \in [0, \frac{\pi}{K})$, $K \geq 2$, let $\Delta \phi$ be the triangle with vertices at $\{\frac{1}{\cos t}, e^{it}, e^{-it}\}$; if for any $\eta \in \mathbb{C}$, $W(A) \subset \phi$, then $W(A^K) \subset \eta^K \Delta(K\phi)$.

2.3 Rational Functions

The results in the two previous sections are in some sense local and are independent of the behavior of f in the entire plane. A natural extension to a global mapping theorem would be that of a rational function. The following theorem provides a mapping for rational functions using the concept of the convex kernel of a set. The *kernel* K of a set S is the subset of S with respect to which it is star-shaped.

Theorem 2.3-1. *Let f be a rational function with $f(\infty) = \infty$. For any compact convex set F in $\mathbb{C}$, let $E = f^{-1}(F)$ and K be the convex kernel of E. Then, for any operator A, with $\overline{W(A)} \subset \operatorname{Int} K$, we have $W(f(A)) \subset F$.*

Proof. Let ℓ be a line of support and tangent to F at b. We shall prove that $W(f(A))$ is contained in the closed half-plane bounded by ℓ and containing F. Let θ be the angle between the real axis and ℓ oriented in such a way that F lies to the left of ℓ. We can assume that b is not a branch point of f^{-1} as there are only a finite number of them. The poles of $[f(z)-b]^{-1}$ are all simple, and we can use the partial fraction expansion (as $f(\infty) = \infty$)

$$[f(z) - b]^{-1} = \sum_i f'(C_i)(z - C_i)^{-1}, \quad \text{with} \quad f'(C_i) \neq 0.$$

Replacing z by A,

$$[f(A) - b]^{-1} = \sum_i f'(C_i)^{-1}(A - C_i I)^{-1}.$$

We thus have

$$\text{Im}\,[e^{i\theta}(f(A) - b)^{-1}] = \sum e^{i\theta} f'(C_i)^{-1}(A - C_k I)^{-1}.$$

We will prove that for each i

(2.3-1) $$\text{Im}\,[e^{-i\theta} f'(C_i)(A - C_i I)] \geq 0.$$

This, in turn, implies that

(2.3-2) $$\text{Im}\,[e^{-i\theta}(f(A) - bI)] \geq 0$$

since, for any invertible operator T, $\text{Im}\,T \geq 0$ is equivalent to $\text{Im}\,T^{-1} \leq 0$.

At any point C_i, we have $f(C_i) = b$ and $f'(C_i) \neq 0$. Thus, there is a neighborhood of C_i on which f is conformal and maps it into a neighborhood of b. Since $b \in \partial F$, we have $C_i \in \partial E$, which has a tangent ℓ_i at C_i. If θ_i is the inclination of L_i, we have $e^{-i\theta} f'(C_i) = e^{-i\theta_i}|f'(C_i)|$. Since $W(A)$, in the half-plane left of ℓ_i, we have

$$\text{Im}\,[e^{-i\theta_i}(A - C_i I)] \geq 0,$$

and hence (2.3-1) holds.

We can easily see that, in the above argument, the assumption that ℓ is a line of support and a tangent is sufficiently general, since the exceptions are at most countable. Further, if $\text{Int}\,K$ is empty, there is nothing to be proved. If K is contained in a straight line, then A is normal and

(2.3-3) $$W(f(A)) = \text{co}\,\sigma(f(A)) = \text{co}\,f(\sigma(A)) \subset \text{co}\,f(W(A)).$$

Since $f(K) \subset F$ and F is convex, we have $W(f(A)) \subset F$. □

We remark that the condition $\overline{W(A)} \subset \text{Int}\,K$ can be modified to $W(A) \subset K$ by using the continuity of the numerical range.

Notes and References for Section 2.3

The important theorem on mapping rational functions is due to

T. Kato (1965). "Some Mapping Theorems for the Numerical Range," *Proc. Japan Acad.* **41**, 652–655.

There is some gap between Section 2.2 and Section 2.3. For example, in the paper just cited, it is shown that for $f(z)$ a function analytic on the half-plane $P = \{z : \operatorname{Re} z \geq 0\}$ and A a bounded operator with numerical range $W(A)$ contained in P, then $W(f(A))$ is contained in the closed convex hull of $f(P)$. But one cannot generate half-plane numerical range mapping theorems automatically from disk mapping theorems because linear fractional transformations such as $f(z) = (1-z)(1+z)^{-1}$, which map D onto P, do not map the numerical range in the same way.

A related gap is the extension of numerical range mapping theorems from the A bounded to the A unbounded case. For example, the Cayley transform from A unbounded symmetric to $V = (A-iI)(A+iI)^{-1}$ bounded isometric will not carry the numerical ranges in the same way.

2.4 Operator Products

We saw earlier that the numerical radius behaves better than the numerical range under mappings. Let us now consider the two basic algebraic operations of linear structure, addition and multiplication. Because the numerical range is subadditive,

$$W(A+B) \subset W(A) + W(B),$$

it is often relatively easy to determine inclusion sets for the numerical range of a sum. As the example given at the beginning of this chapter shows, general results for $W(AB)$ are harder. For example, the mapping theorems in the previous sections apparently are not very applicable to the multiplicative situation. In the next chapter, we will develop methods that are more natural to it.

In the next section, some results for $W(AB)$ will be obtained for A and B commuting selfadjoint or normal operators. In the present section, we are able to obtain some information about the spectrum of a product AB from the numerical ranges $W(A)$ and $W(B)$.

Theorem 2.4-1. *If $0 \notin \overline{W(A)}$, then*

(2.4-1) $$\sigma(A^{-1}B) \subset \frac{\overline{W(B)}}{\overline{W(A)}} = \left\{ \frac{\lambda}{\mu}, \ \lambda \in \overline{W(B)}, \ \mu \in \overline{W(A)} \right\}.$$

Proof. Since $0 \notin \overline{W(A)}$ and $\sigma(A) \subset \overline{W(A)}$, we see that A is invertible. Now let $\lambda \in \sigma(A^{-1}B)$. Hence, $0 \in \sigma(A^{-1}B - \lambda I)$. However,

$$A^{-1}B - \lambda I = A^{-1}(B - \lambda A).$$

Hence, if $0 \in \sigma(A^{-1}B)$, then $0 \in \sigma(B - \lambda A)$. Thus $0 \in \overline{W(B - \lambda A)}$ or equivalently $\lambda \in \frac{\overline{W(B)}}{\overline{W(A)}}$. □

Theorem 2.4-1 has a number of applications. Let us note three of them here.

An operator is called positively stable if all the elements of its spectrum are in the right half-plane. In particular, a selfadjoint operator is stable if it is positive definite.

Theorem 2.4-2. *Let P be positive definite and A stable. Then PA is stable.*

Proof. By Theorem 2.4-1, we have

$$(2.4\text{-}2) \qquad \sigma(PA) = \sigma[(P^{-1})^{-1}A] \subset \frac{\overline{W(A)}}{\overline{W(P^{-1})}}.$$

Since $\overline{W(P^{-1})} = \overline{W(P)}^{-1}$ has only positive real elements and $W(A)$ has elements with positive real parts, we conclude that PA is stable. Note that P and A are invertible and so is PA. □

To study another application, recall that any operator T has a polar decomposition $T = UP$. If T is invertible, then U is unitary and P is positive definite. A unitary operator U is said to be *cramped* if its spectrum is contained in an arc of the unit circle with central angle less than π.

Theorem 2.4-3. *If $0 \notin \overline{W(T)}$, then the unitary part of T is cramped.*

Proof. Let $T = UP$. $0 \notin \overline{W(T)}$ implies that the convex set $W(T)$ is contained in a sector

$$(2.4\text{-}3) \qquad S = \{re^{i\theta} : r > 0, \; \theta_1 \leq \theta \leq \theta_2\},$$

where the difference $\theta_2 - \theta_1 < \pi$.

So, $U = TP^{-1}$, and since $0 \notin \overline{W(T^{-1})}$, using Theorem 2.4-1, we have

$$\sigma(U) \subset \frac{\overline{W(P^{-1})}}{\overline{W(T^{-1})}}.$$

Notice that $W(P^{-1})$ is a closed interval on $\mathbb{R}$. If $z \in W(T^{-1})$, we have $z = \langle T^{-1}x, x \rangle$ for some x, $\|x\| = 1$. So $z = \langle y, Ty \rangle$, where $y = Tx$ and $\arg \langle y, Ty \rangle = -\arg \langle Ty, y \rangle$. Hence, $\frac{W(P^{-1})}{W(T^{-1})}$ is contained in the sector $\theta_1 \leq \theta \leq \theta_2$. So $\sigma(U)$ is contained in the set $\{e^{i\theta} : \theta_1 \leq \theta \leq \theta_2\}$. □

In the case of unitary A and B we can determine $\overline{W(AB)}$ exactly as follows. The assumption is that $0 \notin W(A) \cap W(B)$, which is equivalent to assuming that one of the operators has a spectrum of arc length less than π, i.e., that one of the operators has cramped spectrum.

Theorem 2.4-4. Let A and B be unitary operators on a Hilbert space such that $\sigma(A) \subset \text{arc}\,\Gamma_A$ and $\sigma(B) \subset \text{arc}\,\Gamma_B$, where either arc length $\Gamma_A < \pi$ or arc length $\Gamma_B < \pi$. Then

(2.4-4) $$W(AB) \subset \text{cl conv hull}\,(\Gamma_A \Gamma_B).$$

Proof. Here $\Gamma_A \Gamma_B$ denotes the set of points $z_A z_B$, $z_A \in \Gamma_A$, $z_B \in \Gamma_B$. We need consider only the case arc length $\Gamma_A < \pi$ because A and B are interchangeable, due to the fact that $\Gamma_A \Gamma_B = \Gamma_B \Gamma_A$.

Since A is unitary, we have $\overline{W(A)} \subset \text{cl conv hull}\,(\Gamma_A)$; thus arc length $\Gamma_A < \pi \Rightarrow 0 \notin \overline{W(A)} \Rightarrow 0 \notin \overline{W(A^{-1})}$. By Theorem 2.4-1, on the spectrum of a product we have $\sigma(AB) \subset \overline{W(B)}/\overline{W(A^*)} \subset S$, where S denotes the sector $\alpha \leq \theta \leq \beta$, where $\alpha = \inf\{\theta_1 + \theta_2 \mid e^{i\theta_1} \in \Gamma_A, e^{i\theta_2} \in \Gamma_B\}$ and where β is the supremum of the same set. Since AB is unitary, its numerical range closure is the convex hull of its spectrum, and hence

$$\overline{W(AB)} \subset \text{cl conv hull}\,(\Gamma_A \Gamma_B). \quad \square$$

Notes and References for Section 2.4

Theorem 2.4-1 was shown by
J. P. Williams (1967). "Spectra of Products and Numerical Ranges," *J. Math. Anal. Appl.* **17**, 214–220.
Earlier versions for matrices were given by
H. Wielandt (1951). *National Bureau of Standards Report* 1367, Washington, D.C.
Versions of Theorems 2.4-3 and 2.4-4 were given by
S. Berberian (1964). "The Numerical Range of a Normal Operator," *Duke Math. J.* **31**, 479–483.

2.5 Commuting Operators

In this section, we will present some results concerning $W(AB)$ and $w(AB)$ when $AB = BA$. As the example at the beginning of this chapter shows, not much can be expected generally of the numerical range $W(AB)$. However, in some special instances, one can say something such as the following, for example,

Theorem 2.5-1. Let A be a nonnegative, selfadjoint operator and $AB = BA$. Then $W(AB) \subset W(A)W(B)$.

Proof. $\langle ABx, x \rangle = \langle BA^{1/2}x, A^{1/2}x \rangle$, where $A^{1/2}$ is the nonnegative square root of A. Thus $\langle ABx, x \rangle = \langle Bg, g \rangle \|A^{1/2}x\|^2 = \langle Bg, g \rangle \langle Ax, x \rangle$, where $g = \frac{A^{1/2}x}{\|A^{1/2}x\|}$ with $A^{1/2}x \neq 0$. $\square$

2.5 Commuting Operators

It turns out that more can be expected for the numerical radius $w(AB)$, in view of the power inequality, Theorem 2.1-1.

First, let us note some readily available facts. Simple examples show that $w(AB)$ can exceed the product $w(A)w(B)$. These examples show that this can be the case even when A and B commute. For example, take the right shift

$$A = \begin{bmatrix} 0 & 0 & 0 & 0 \\ 1 & 0 & 0 & 0 \\ 0 & 1 & 0 & 0 \\ 0 & 0 & 1 & 0 \end{bmatrix}.$$

Then by (1.3-4), we know that $w(A) = \cos(\pi/5) = 0.80901699$. On the other hand, straightforward computation (similar to that of the first example in the book) shows that $w(A^2) = w(A^3) = 0.5$, so that

$$0.5 = w(A \cdot A^2) > w(A) \cdot w(A^2) = 0.4045085.$$

In the affirmative direction, we easily have the following results.

Theorem 2.5-2. *It is always the case that*

$$w(AB) \leq 4w(A)w(B).$$

When $AB = BA$, it always holds that

$$w(AB) \leq 2w(A)w(B).$$

Proof. From norm equivalence, Theorem 1.3-1, we know that

$$w(AB) \leq \|A\|\|B\| \leq 4w(A)w(B).$$

In the commuting case, we may assume $w(A) = w(B) = 1$ and show that $w(AB) \leq 2$. Because of Theorem 1.3-1, the numerical radius is a norm. Hence, by the triangle inequality, the power inequality theorem (Theorem 2.1-1), and the subadditivity of w, we have

(2.5-1)
$$\begin{aligned} w(AB) &\equiv w\left(\frac{1}{4}[(A+B)^2 - (A-B)^2]\right) \\ &\leq \frac{1}{4}[w((A+B)^2) + w((A-B)^2)] \\ &\leq \frac{1}{4}[(w(A+B))^2 + (w(A-B))^2] \\ &\leq \frac{1}{4}[(w(A)+w(B))^2 + (w(A)+w(B))^2] \\ &= 2. \quad \square \end{aligned}$$

Next, we turn to the case in which A and B *double commute*: $AB = BA$ and $AB^* = B^*A$. Note that by taking adjoints, B and A also double commute. Under this assumption, one can get closer to the elusive $w(AB) \leq$

2. Mapping Theorems

$w(A)w(B)$ situation, namely, one can show that $w(AB) \leq w(A)\|B\|$. To prove this, we first arrange two lemmas.

Lemma 2.5-3. *Let A be a unitary operator that commutes with another operator B. Then $w(AB) \leq w(B)$.*

Proof. Assuming that $w(B) = 1$, we have

$$\langle (I - zB)f, f \rangle \geq 0, \quad |z| \leq 1.$$

In particular,

$$\langle (I - e^{i\theta}B)f, f \rangle \geq 0 \quad \text{for} \quad \theta \in [0, 2\pi].$$

Therefore,

(2.5-2) $$\int_0^{2\pi} \langle (1 - e^{i\theta}B)f, f \rangle dE_\theta \geq 0,$$

where $\{E_\theta\}$ is the spectral family for A over the segment $0 \leq \theta \leq 2\pi$. The family E_θ can be approximated by a sequence of polynomials in A and A^*. Since A and A^* commute with B and the integral, and the inner products are continuous with respect to their arguments, we have

(2.5-3) $$0 \leq \int_0^{2\pi} \langle (I - e^{i\theta}B)f, f \rangle dE_\theta = \langle (I - AB)f, f \rangle. \quad \square$$

Lemma 2.5-4. *Let A be an isometry and $AB = BA$. Then $w(AB) \leq w(B)$.*

Proof. $A^*A = I$ and hence $\langle ABf, f \rangle = \langle A^*AABf, f \rangle = \langle ABAf, Af \rangle$. So we need to consider only the restrictions to $R(A)$, which is closed. Notice that on the range of A, A is unitary, because $A^*A = I$ and for any $f = Ag \in R(A)$, we have $AA^*f = AA^*Ag = Ag = f$. So $AA^* = I$ on $R(A)$. Since A is unitary on $R(A)$, by Lemma 2.5-3 we have $w(AB) \leq w(B)$ on $R(A)$. $\square$

Theorem 2.5-5 (Double commute). *If the operators A and B double commute, then $w(AB) \leq w(B)\|A\|$.*

Proof. Without loss of generality, we may scale A so that $\|A\| < 1$. Then, it is enough to show that $w(AB) \leq w(B)$. Let $D = (I - A^*A)^{1/2}$ be the positive square root, and consider the two operators S and T on $K = H \oplus H \oplus H \oplus \cdots$ defined by

(2.5-4) $$S(f_1, f_2, f_3, \ldots) = (Af_1, Df_1, f_2, f_3, \ldots),$$
$$T(f_1, f_2, f_3, \ldots) = (Bf_1, DBD^{-1}f_2, DBD^{-1}f_3, \ldots).$$

Then S and T commute and S is an isometry, the former because D^{-1} commutes with B, the latter from using $D^2 = I - A^*A$ in
$$\|Sf\|_K^2 = \|Af_1\|^2 + \|Df_1\|^2 + \|f\|_K^2 - \|f_1\|^2 = \|f\|_K^2.$$
By Lemma 2.5-4, we therefore have $w_K(ST) \leq w_K(T)$ for the dilations S and T of A and B. Since the numerical range of a direct sum operator is the convex hull of the summand's numerical ranges, we have $w(AB) \leq w_K(ST)$, ST being a dilation of AB. Moreover, $w_K(T) = \max\{w(B), w(DBD^{-1})\} = w(B)$. □

Corollary 2.5-6. Let A be a normal operator commuting with B. Then $w(AB) \leq w(A)w(B)$.

Proof. By Fuglede's theorem (see [H]), A commutes with B and B^*, and $\|A\| = w(A)$ when A is normal. □

The following result covers some operators that are not normal but are square roots of a scalar; for example, the matrix
$$\begin{bmatrix} 0 & a \\ b & 0 \end{bmatrix}.$$

Theorem 2.5-7. Let $AB = BA$ and $A^2 = \alpha I$, where $\alpha \in \mathbb{C}$. Then $w(AB) \leq w(B)\|A\|$.

Proof. Let H_1 be the Hilbert space H renormed by $\langle f, g \rangle_1 = \langle Df, Dg \rangle$, where
$$D^2 = I + A^*A + A^{*2}A^2 + \cdots + (A^*)^n A^n + \cdots,$$
where we assume $\|A\| < 1$. Let $K = H_1 \oplus H_1 \oplus \cdots$, and define the operators S and T on K by

(2.5-5) $\qquad S(f_1, f_2, \ldots) = (Af_1, D^{-1}f_1, f_2, f_3, \ldots)$

and

(2.5-6) $\qquad T(f_1, f_2, \ldots) = (Bf_1, D^{-1}BDf_2, D^{-1}BDf_3, \ldots).$

Notice that $ST = TS$ and $w_K(T) = w(B)$, so if $f = (f_1, f_2, \ldots)$, we have
$$\langle STf, f \rangle_k = \langle D^2 ABf_1, f_1 \rangle + \langle Bf_1, Df_2 \rangle + \langle BDf_2, Df_3 \rangle + \cdots.$$
Choosing $Df_2 = (I - A^*A + (A^*)^2 A^2 - (A^*)^3 A^3 + \cdots)f_1$, we observe that

(2.5-7) $\qquad \langle [I + A^*A + (A^*)^2 A^2 + \cdots] ABf_1, f_1 \rangle$
$\qquad \leq \langle [I + (A^*A + (A^*)^2 A^2 + \cdots] f_1, f_1 \rangle.$

Since $A^2 = \alpha I$, we have
$$\langle [1 + |c|^2 + |c|^4 + \cdots] ABf_1, f_1 \rangle \leq \langle [1 + |c|^2 + |c|^4 + \cdots] f_1, f_1 \rangle.$$ □

Notes and References for Section 2.5

In the commuting case it was observed by
K. Gustafson (1968). "The Angle of an Operator and Positive Operator Products," *Bull. Amer. Math. Soc.* **74**, 488–492.
that for positive selfadjoint A and bounded accretive B, BA is accretive. Later this was extended to $W(AB) \subset W(A)W(B)$ for positive selfadjoint A and bounded commuting A and B, in
R. Bouldin (1970). "The Numerical Range of a Product," *J. Math. Anal. Appl.* **32**, 459–467.
These commuting results are rather immediate by use of $A^{\frac{1}{2}}$.

Theorem 2.5-2 for commuting A and B was observed in
J. Holbrook (1969). "Multiplicative Properties of the Numerical Radius in Operator Theory," *J. Reine Angew. Math.* **237**, 166–174.
This was the most important paper for these commuting operator numerical radius questions. In particular, not only the first but three proofs of the double commute result (Theorem 2.5-5) were given there. Two of these proofs used dilation theory (see the next section).

Theorem 2.5-5 was extended by
R. Bouldin (1971). "The Numerical Range of a Product, II," *J. Math. Anal. Applic.* **33**, 212–219.
Lemma 2.5-4 is obtained there as well, its proof depending on the dilation theory. Our proof of Lemma 2.5-4 does not need the dilation methods. However, our proof of Theorem 2.5-5 is essentially that of the just cited paper.

Historically, Corollary 2.5-6 preceded Theorem 2.5-5 and is easily shown by use of the spectral theorem for normal operators. More generally, we may say from Theorem 2.5-5 that the sought inequality $w(AB) \leqq w(A)w(B)$ holds for all double commuting A and B whenever A is also *normaloid*: $w(A) = \|A\|$. Normaloid operators will be discussed further in Chapter 6.

Theorem 2.5-7 was given by
D. K. Rao (1994). "Rango Numerico de Operadores Conmutativos," *Revista Colombiana de Matematicas* **27**, 231–233.

The question of whether, when A and B commute,

$$(2.5\text{-}8) \qquad w(AB) \leqq w(A)\|B\|$$

was open for about twenty years. Its falsehood was finally resolved by counterexample in
V. Müller (1988). "The Numerical Radius of a Commuting Product," *Michigan Math. J.* **35**, 255–260.
Müller's approach was computational, and the counterexample was found in a 12-dimensional Hilbert space. The counterexample can even be constructed with B a polynomial in A.

The related question of the best constants for the inequality

(2.5-9) $$w(AB) \leqq Cw(A)\|B\|$$

for commuting A and B has also been considered; see
K. Okubo and T. Ando (1976). "Operator Radii of Commuting Products," *Proc. Amer. Math. Soc.* **56**, 203–210.
K. Davidson and J. Holbrook (1988). "Numerical Radii of Zero-One Matrices," *Michigan Math. J.* **35**, 261–267.

The last cited paper also gives a simple counterexample to the inequality (2.5-8). Let

$$S = S_9 = \begin{bmatrix} 0 & 1 & & & 0 \\ & \ddots & & & \\ & & & & \\ & & & \ddots & 1 \\ 0 & & & & 0 \end{bmatrix}$$

be the left shift in $\mathbb{C}^9$, and let

$$T = S^3 + S^7 = \begin{bmatrix} 0 & 0 & 0 & 1 & 0 & 0 & 0 & 1 & 0 \\ & & & & 1 & & & & 1 \\ & & & & & 1 & & & 0 \\ & & & & & & 1 & & 0 \\ & & & & & & & 1 & 0 \\ & & & & & & & & 1 \\ & & & & & & & & 0 \\ & & & & & & & & 0 \\ & & & & & & & & 0 \end{bmatrix},$$

so that

(2.5-10) $$TS = S^4 + S^8 = \begin{bmatrix} 0 & 0 & 0 & 0 & 1 & 0 & 0 & 0 & 1 \\ & & & & & 1 & & & 0 \\ & & & & & & 1 & & 0 \\ & & & & & & & 1 & 0 \\ & & & & & & & & 1 \\ & & & & & & & & 0 \\ & & & & & & & & 0 \\ & & & & & & & & 0 \\ & & & & & & & & 0 \end{bmatrix}.$$

Clearly, $TS = ST$. By (1.3-4) we know that $w(S) = \cos(\pi/10) = 0.95105652$. It can be shown (see Section 5.3) that $w(T)$ has the same value, but $w(TS) = 1$. Since $\|S\| = 1$, the constant in (2.5-9) for this example is $c = 1.0514622$.

2.6 Dilation Theory

Because the dilation theory of Sz.-Nagy, Foias, Berger, Halmos, Ando, and others has impacted the numerical range theory, we want to explain some of that theory briefly in this section. Recall that some of the proofs in Section 2.1 and Section 2.5 employed dilations. There are many dilation theories (see the Notes). Here we will describe only those based on extending the Hilbert space by direct sums.

Let us consider the matrix case first. For $n \times n$ matrices A and B, the direct sum is the $2n \times 2n$ matrix

$$A \oplus B = \begin{bmatrix} A & 0 \\ 0 & B \end{bmatrix}$$

on $H \oplus H$. By the *direct sum property* (easily verified; see [H]),

(2.6-1) $\qquad W(A \oplus B) = \text{conv hull}(W(A) \cup W(B)),$

one immediately sees many opportunities for constructing and manipulating numerical ranges of A extended to direct sums, finite or infinite.

Matrix dilations are a generalization of direct sums and, in connection with numerical ranges, open up even further opportunities for numerical range constructions. For any $n \times m$ matrix, any strictly larger matrix

$$\widetilde{A} = \begin{bmatrix} A & B \\ C & D \end{bmatrix}$$

is called a dilation of A to a larger space. For example, for any $n \times n$ A, one can always get a small $(2n \times 2n)$ normal dilation; for example,

(2.6-2) $\qquad \widetilde{A} = \begin{bmatrix} A & A^* \\ A^* & A \end{bmatrix}.$

By the *submatrix inclusion property* (this is easily verified; see Theorem 5.1-2 or [M]), $W(A) \subset W(\widetilde{A})$ for any matrix dilation $\widetilde{A}$. It is relatively easy to show (see, e.g., [HJ2]) that

(2.6-3) $\qquad W(A) = \bigcap W(\widetilde{A})$

over all $2n \times 2n$ normal dilations of A.

Hilbert space dilations and their use in numerical range theory generalize these ideas. Any extension $\widetilde{T}$ of an operator T to a Hilbert space K strictly containing H such that

(2.6-4) $\qquad T = P_H \widetilde{T} \quad \text{on} \quad H,$

where P_H denotes the orthogonal projection of K onto H, is called a *dilation* of T. It becomes a *strong dilation* if

(2.6-5) $\qquad T^n = P_H \widetilde{T}^n \quad \text{on} \quad H$

for all positive integers n. If $\widetilde{T}$ is unitary and

(2.6-6) $$T^n = \rho P_H \widetilde{T}^n \quad \text{on} \quad H$$

it is called a *strong ρ unitary dilation*.

For arbitrary contractions ($\|T\| \leq 1$), and all bounded operators T can be made contractions by normalizing them, one may get a rather small $(H \oplus H)$ unitary dilation of T, namely

(2.6-7) $$\widetilde{T} = \begin{bmatrix} T & (I - TT^*)^{\frac{1}{2}} \\ (I - T^*T)^{\frac{1}{2}} & -T^* \end{bmatrix},$$

where the off-diagonal operators are the unique selfadjoint positive semi-definite square roots guaranteed by the nonnegativity

(2.6-8) $$\langle (I - TT^*)x, x \rangle = \|x\|^2 - \|T^*x\|^2 \geq 0.$$

The dilation (2.6-7) turns out to be useful, even though it does not by itself have the operator power properties desired of a strong dilation. To note the latter, take the trivial case of T selfadjoint; then $\widetilde{T}^2 = I$! The key to the unitarity of the dilation (2.6-7) is the block intertwining property

(2.6-9) $$(I - TT^*)^{\frac{1}{2}} T = T(I - T^*T)^{1/2},$$

which follows from the identity $T(I - T^*T) = (I - TT^*)T$.

A main result in the Hilbert space dilation theory is the following.

Theorem 2.6-1. *For every contraction T on a Hilbert space, there exists a unitary dilation U on some Hilbert space K containing H as a subspace such that*

$$T^n x = P_H U^n x, \quad n = 0, 1, 2, 3, \ldots$$

One can require that U be minimal, in the (cyclic) sense that

$$\bigcup_{n=-\infty}^{n=\infty} U^n(H) = K.$$

U is then determined uniquely, up to isometric isomorphism.

Proof. Several proofs have been given (see the Notes). The first started with the fact that T contractive implies the positivity

(2.6-10) $$\operatorname{Re}\left[\langle x, x \rangle + 2 \sum_{n=1}^{\infty} \lambda^n \langle T^n x, x \rangle\right] \geq 0$$

for all $|\lambda| < 1$. From this the Stieltjes integral representation of analytic functions with nonnegative real part in the unit circle leads to the representation

(2.6-11) $$T^n = \int_0^{2\pi} e^{in\theta} d\mu_\theta,$$

44   2. Mapping Theorems

where μ_θ is a (obtained as a boundary value limit) measure on $[0, 2\pi]$. By a result of Neumark, this measure may be "dilated" to a unitary spectral measure E, i.e., $\mu(\mathcal{B})x = P_H E(\mathcal{B})x$ for all Borel sets $\mathcal{B}$, and from this one takes the unitary dilation of T to be $U = \int_0^{2\pi} e^{i\theta} dE_\theta$. Later proofs utilized a construction like (2.6-7) placed within a setting of an infinite direct sum of copies of H and coupled to some shifts therein, similar to the constructions used in Section 2.5. □

The ρ-dilations are the most interesting ones for use in connection with the numerical range $W(T)$. To motivate this, note that the rather easy matrix normal dilation (Eq. (2.6-2)) has square

$$(\widetilde{A})^2 = \begin{bmatrix} A^2 + (A^*)^2 & AA^* + A^*A \\ A^*A + AA^* & (A^*)^2 + A^2 \end{bmatrix},$$

and in the trivial case where A is symmetric, we see that $A^2 = \frac{1}{2} P_H(\widetilde{P})^2$ on H. The idea is to go the other way, to push ρ to as high a value as possible. This is the meaning of the dilation used in the proof of the power inequality in Section 2.1, that from numerical radius $w(T) \leqq 1$ we are guaranteed a unitary dilation $\widetilde{T}$ of T such that $T^2 = 2P_H\widetilde{T}^2$. In other words, we want the dilation to have good numerical range and numerical radius properties and still, when projected, we want it to "shrink" T and its powers T^n as much as possible. It is easy to see from the basic relation $w(T) \leqq \|T\| \leqq 2w(T)$ that 2 is the sharpest ρ that can be expected. For the power inequality, the strong dilation must achieve 2 exactly.

Notes and References for Section 2.6

The first step in the dilation theory seems to have been that of
P. Halmos (1950). "Normal Dilations and Extensions of Operators," *Summa Brasil. Math.* **2**, 125–134.
where the dilation (2.6-7) was given. Theorem 2.6-1 was proved by
B. Sz.-Nagy (1953). "Sur les Contractions de l'espace de Hilbert," *Acta Sci. Math.* **15**, 87–92.
This was the first general operator strong dilation result and had a substantial impact on the harmonic analysis of operators by use of extensions of them to larger Hilbert spaces. The term "harmonic analysis" is appropriate in terms of the proof of Theorem 2.6-1 we sketched, as (2.6-11) may be regarded as a "Fourier coefficient" of the measure μ_θ on the unit circle. The full theory of such dilations may be found in the book
B. Sz.-Nagy and C. Foias (1970). *Harmonic Analysis of Operators on Hilbert Space*, North Holland.

At the end of the proof of Theorem 2.6-1 we mentioned a later proof that brings in Halmos' dilation construction (2.6-7). This was given by
J. J. Schäffer (1955). "On Unitary Dilations of Contractions," *Proc. Amer. Math. Soc.* **6**, 322.

Schäffer's proof is noteworthy for its brevity, but carries the advantage of directly exhibiting the strong dilation.

Next came a question of whether two commuting contractions T_1 and T_2 could have commuting strong unitary dilations. This was proved by

T. Ando (1961). "On a Pair of Commutative Contractions," *Acta Sci. Math.* **24**, 88–90.

The result does not extend to three commuting operators.

It is known (generalized Wold decomposition; see, e.g., [H]) that every contraction T may be uniquely decomposed as the direct sum of a unitary part and a completely nonunitary part. The latter has a minimal dilation, which is the skew sum of two bilateral shifts. There are connections between this theory and the incoming/outgoing subspaces decompositions of scattering theory, see

P. Lax and R. S. Phillips (1967). *Scattering Theory*, Academic Press, New York.

Indeed, group representation theory is an earlier "dilation" theory, but the purposes and results are quite different.

Just as parts of operator theory may be somewhat irreverently viewed as extending complex numbers $z = e^{i\theta}r$ to operators $T = U|T|$, both written here in their polar forms, where $|T| = (T^*T)^{1/2}$ and U is the partial isometry mapping the range of $|T|$ to the range of T, some of the dilation theory, especially that originated by Sz.-Nagy and Foias, has some motivation in the Hardy space theory of decompositions of a function h (in applications, in the frequency domain) as $h = io$, where i is an inner function and o is an outer function. Remember that inner functions may be thought of as Blaschke products, which absorb all upper half-plane zeros from h. For our own version of this, see

R. Goodrich and K. Gustafson (1981). "Weighted Trigonometric Approximation and Inner–Outer Functions on Higher Dimensional Euclidean Spaces," *J. Approx. Theory* **31**, 368–382.

R. Goodrich and K. Gustafson (1986). "Spectral Approximation," *J. Approx. Theory* **48**, 272–293.

One serious deficiency of the Hilbert space operator dilation theories we have described is that in certain concrete applications in specific function space settings, they will not carry into the dilations needed lattice properties (e.g., preservation of positivity of functions representing probabilities). To that end, another kind of dilation theory was started by

M. Akcoglu (1975). "Positive Contractions on L_1-Spaces," *Math. Zeit.* **143**, 1–13.

For a good treatment of that dilation theory, see

M. Kern, R. Nagel, and G. Palm (1977). "Dilations of Positive Operators: Construction and Ergodic Theory," *Math. Zeit.* **156**, 265–267.

In a recent paper, we employ the Akcoglu dilation theory

I. Antoniou and K. Gustafson (1996). "Dilation of Markov Processes to Dynamical Systems," preprint.

to answer a question of dilating a probabilistic Markov semigroup M_t to a deterministic evolution in a larger space.

Although the Sz.-Nagy–Foias et al. Hilbert space operator dilates a continuous positive family W_t of operators to a unitary family U_t on a larger space, and although the dilations there can always in fact be given a positivity structure, we did not see how to do that in such a way as to accommodate also a meaningful extension of the underlying phase space dynamics. The Akcoglu dilation construction often can be shown to preserve phase space dynamical characteristics, such as ergodicity, and in our case was useful to distinguish physical reversibility and irreversibility.

A third dilation theory is that of Kolmogorov–Rokhlin; see
I. P. Cornfeld, S. V. Fomin, and Ya G. Sinai (1982). *Ergodic Theory*, Springer, New York.
V. Rokhlin (1964). "Exact Endomorphisms of a Lebesgue Space," *Amer. Math. Soc. Transl.* **39**, 1–36.
In this theory, one is able to construct "natural extensions" to what are called exact dynamical systems, which are roughly those that preserve a measure on the phase space dynamics. We employed this dilation theory to show that certain probabilistic dynamical systems may be embedded in a larger deterministic dynamics of what are called Kolmogorov systems, in
I. Antoniou and K. Gustafson (1993). "From Probabilistic Descriptions to Deterministic Dynamics," *Physica A* **197**, 153–166.

One may usefully relate these three dilation theories intuitively as follows. Picture a trajectory dynamics S_t on subsets β of a phase space Ω, and consider the $\mathcal{L}^p(\Omega, \beta, \mu)$ space of functions f (e.g., probability densities) defined over Ω. Then the Sz.-Nagy–Foias et al. theory dilates the functions f; the Kolmogorov–Rokhlin et al. theory dilates the trajectory dynamics S_t; and the Akcoglu–Suchestcon et al. theory dilates the measure μ. All induce dilations of the corresponding operator semigroups W_t.

Endnotes for Chapter 2

One remarkable fact about mapping theorems for the numerical range is that there are at least some of them. For example, the simpler question of just polynomial mapping theorems just for the numerical radius of just 2×2 matrices with real determinant and real trace shows that even the 2×2 case can be quite complicated; see
C. R. Johnson, I. M. Spitkovsky, and S. Gottlieb (1994). "Inequalities Involving the Numerical Radius," *Linear and Multilinear Algebra* **37**, 13–24.
As mentioned there, even the question of when $w(f(A)g(A)) \leqq w(f(A)) w(g(A))$ holds for A nonnormal, even for simple real polynomials such as $f(z) = z$ and $g(z) = z^2$, is unresolved. The case of general matrix products can therefore be expected to be much more difficult.

A natural approach to some of these questions, as we have seen, is through dilations. One might hope in the commuting case that the dilations of S, $\|S\| \leq 1$, and T, with $w(T) \leq 1$, could be combined. However, it was shown by Davidson and Holbrook (see Section 2.5; Notes and References) that for dilation constant $\rho > 1$ there are commuting operators S and T such that $w_\rho(ST) > w_\rho(T)\|S\|$. Here the (homogeneous) operator radii $w_\rho(T)$ are defined by $w_\rho(T) \leqq 1$ if and only if $T \in \mathcal{C}_\rho$, the Sz.-Nagy–Foias class of operators having strong ρ unitary dilation; alternately

$$w_\rho(T) = \inf\{\gamma \mid \gamma > 0,\ \gamma^{-1}T \in \mathcal{C}_\rho\}.$$

These interesting operator radii contain $\|T\| = w_1(T), w(T) = w_2(T)$, and $r(T) = \lim_{\rho \to \infty} w_\rho(T)$. See
J. A. R. Holbrook (1968). "On the power-bounded operators of Sz.-Nagy and Foias," *Acta Sci. Math.* **29**, 299–310.
K. Okubo and T. Ando (1976). "Operator radii of commuting products," *Proc. Amer. Math. Soc.* **56**, 203–210.
The only pair of dilations that can be combined (because the unitary parts commute) are the 1–1 Ando dilations (see Section 2.6; Notes and References).

It would be desirable to have more $W(A)$ mapping theorems. Even "semiglobal" results for sectors and half-planes for certain classes of operators A, and mappings ranging from "easy" such as polynomials to "academic" such as the important transcendental functions, should be better understood.

Such mapping theorems could be sought as motivated by very specific applications or situations. For example, it was shown by
K. Gustafson (1969). "Polynomials of Infinitesimal Generators of Contraction Semigroups," *Notices Amer. Math. Soc.* **16**, 767,
see also
K. Gustafson and G. Lumer (1972). "Multiplicative Perturbation of Semigroup Generators," *Pacific J. Math.* **41**, 731–742,
that if A is a (generally, bounded or unbounded) contraction semigroup generator and $p(z)$ is a polynomial such that there exists a z_0, Re $z_0 > 0$, such that the zeros λ_i of $z_0 - p(z)$ all lie in the resolvent set $\rho(A)$, then $p(A)$ will also be a contraction semigroup generator iff $W(p(A))$ is contained within the left half-plane Re $z \leqq 0$. Contraction semigroup generators will be discussed further in the next chapter. The point raised here is that it would be useful to have more understanding of sectorial or half-plane numerical range mapping theorems for particular operator classes.

The potential subtlety of numerical range mapping properties appears also in related questions of monotonicity of functions of matrices. For example, consider two symmetric positive semi-definite matrices A and B such that $A \geqq B$, i.e., $A - B \geqq 0$. Examples may be easily constructed to show that $A^2 - B^2$ need not be positive. However, the fundamental result that $A^p \geqq B^p$ for all $0 < p \leqq 1$ was established by

K. Löwner (1934). "Uber Monotone Matrixfunktionen," *Math. Z.* **38**, 177–216.

The proof requires analytic continuation into the upper half-plane. Simpler proofs have been given for the case $p = 1/2$. For related results see

E. Heinz (1951). "Beiträge zur Störungstheorie der Spektralzerlegung," *Math. Ann.* **123**, 415–438.

C. Davis (1963). "Notions Generalizing Convexity for Function Defined on Spaces of Matrices," in *Convexity* (V. Klee, ed.), Amer. Math. Soc., 187–201.

K. Bhagwat and A. Subramanian (1978). "Inequalities Between Means of Positive Operators," *Math Proc Camb. Phil. Soc.* **83**, 393–401.

Recently, Löwners power inequality has been generalized by

T. Furuta (1987). "$A \geq B \geq 0$ assure $(B^r A^p B^r)^{1/q} \geq B^{(p+2r)/q}$ for $r \geq 0$, $p \geq 0$, $q \geq 1$ with $(1 + 2r)q \geq p + 2r$," *Proc. Amer. Math. Soc.* **101**, 85–88.

T. Furuta (1995). "Extension of the Furuta Inequality and Ando–Hiai Log–Majorization," *Lin. Alg. Appl.* **219**, 139–155.

3

Operator Trigonometry

Introduction

The concept of the *angle* of an operator T was introduced in 1967 for use in a perturbation theory of semigroup generators. From this has developed what we call an *operator trigonometry*, whose theory and applications are still evolving. Its properties are intimately associated with the numerical range $W(T)$ and are described in this chapter. Some applications will be described in the next chapter.

Because we have played a major role in the creation of this theory, and because no comprehensive account of it occurs anywhere else in book form, we will attempt to present all the main currents of its development to date. In the endnotes section we will give a rather complete set of references, with some historical comment.

3.1 Operator Angles

The *cosine* of T in $B(H)$ was originally defined as follows:

$$(3.1\text{-}1) \qquad \cos T = \inf_{x \in D(T)} \frac{\operatorname{Re} \langle Tx, x \rangle}{\|Tx\| \cdot \|x\|}, \qquad x \neq 0, \; Tx \neq 0$$

for arbitrary operators in a Banach semi-innerproduct space. We will restrict attention here primarily to the case of $T \in B(H)$. Clearly, (3.1-1) is a *real* cosine defined for the real part of the numerical range of T. The *total cosine*

$$|\cos|T = \inf_{x \in H} \frac{|\langle Tx, x \rangle|}{\|Tx, x\| \cdot \|x\|}, \qquad x \neq 0, \; Tx \neq 0,$$

and *imaginary cosines* are similarly defined. Most of our own interest originally centered on semigroup theory and in particular on *accretivity* ($\operatorname{Re} \langle Tx, x \rangle \geqq 0$) preserving questions. Thus to some extent we will retain that point of view here. Comparable half-space preserving results may be obtained by easy modification.

50   3. Operator Trigonometry

The expression (3.1-1) defines the angle $\phi(T)$. The angle $\phi(T)$ measures the maximum (real) turning effect of T. Later we will consider some operator subangles $\phi_i(T)$ that measure smaller critical turning effects of the operator T. Mixed trigonometries for, say, two operators A and B can be built from $\phi(A)$ and $\phi(B)$. Further, one could investigate trigonometries of algebras and other more general structures. Clearly, $\phi(T) = \phi(T^{-1}) = \phi(cT) = \phi(cT^{-1})$ for any real scalar multiples $c \neq 0$. The reader should note that $\cos T$ here is a geometrical notion, and not the entity $\cos(T)$ defined as a power series in a functional calculus of T.

A general bound for $\cos T$ is available by elementary considerations.

Theorem 3.1-1. *For any operator T, its cosine is bounded by the upper and lower numerical radii $m(T)$ and $w(T)$:*

$$(3.1\text{-}2) \qquad \frac{m(T)}{\|T\|} \leq |\cos|T \leq \frac{w(T)}{\|T\|}.$$

Proof. Both bounds follow from the elementary fact that for positive quantities a_n, b_n, one has

$$\inf(a_n)\inf(b_n) \leq \inf(a_n b_n) \leq \sup(a_n)\inf(b_n).$$

The same bounds hold for (real) $\cos T$. The upper bound follows from

$$\inf_{\|x\|=1} \frac{\operatorname{Re}\langle Tx, x\rangle}{\|Tx\|} \leq \sup_{\|x\|=1} \operatorname{Re}\langle Tx, x\rangle \cdot \inf_{\|x\|=1} \frac{1}{\|Tx\|},$$

and the lower bound from two infs. □

These bounds are generally not very sharp, although for operators far from normality, they may indeed be sharp. For the important class of selfadjoint operators, the cosine is known.

Theorem 3.1-2. *For T a strongly positive ($m > 0$) selfadjoint operator,*

$$(3.1\text{-}3) \qquad \cos T = \frac{2\sqrt{mM}}{m+M}.$$

where $m = m(T)$ and $M = w(T)$.

Proof. This is most easily seen (in retrospect; see the Notes) by use of the Kantorovich inequality:

$$(3.1\text{-}4) \qquad \max_{\|y\|=1}\{\langle y, Ty\rangle\langle y, T^{-1}y\rangle\} = \frac{1}{4}\left(\sqrt{\frac{M}{m}} + \sqrt{\frac{m}{M}}\right)^2.$$

First, we change the $\cos T$ minimization to a maximization:

$$\left(\min_x \frac{\langle Tx, x\rangle}{\|Tx\| \cdot \|x\|}\right)^{-2} = \left(\max_x \frac{\|Tx\|\|x\|}{\langle Tx, x\rangle}\right)^2 = \max_x \frac{\|Tx\|^2\|x\|^2}{\langle Tx, x\rangle^2}.$$

3.1 Operator Angles   51

In the maximization, we then replace x with $\tilde{x} = \langle Tx, x \rangle^{-1/2} x$ so that

$$\max_x \frac{\|Tx\|^2 \|x\|^2}{\langle Tx, x \rangle^2} = \max_{\langle Tx,x \rangle = 1} \|Tx\|^2 \|x\|^2$$

$$= \max_{\langle Tx,x \rangle = 1} \langle T^{1/2}x, T^{3/2}x \rangle \langle T^{1/2}x, T^{-1/2}x \rangle.$$

Then, we let $y = T^{1/2}x$ and require that $\|y\|^2 = \langle Tx, x \rangle = 1$ so that the maximization just written is that of Kantorovich:

$$\max_{\|y\|=1} \langle y, Ty \rangle \langle y, T^{-1}y \rangle = \frac{1}{4} \left(\frac{M}{m} + \frac{m}{M} + 2 \right) = \frac{M^2 + m^2 + 2mM}{4mM}.$$

Thus, $\cos T$ is the inverse square root of this quantity:

$$\cos T = \frac{2\sqrt{mM}}{M+m}. \qquad \square$$

Although we are chiefly interested in bounded operators $B(H)$ in this book, it is worth noting that for unbounded operators in a Hilbert space, $\cos T$ is always 0. This fact geometrically distinguishes the topological notions of boundedness and unboundedness for accretive operators. In other words, an unbounded operator has no restriction on its turning rates and may achieve turning angles arbitrarily close to 90° regardless of its particular operator structure.

Theorem 3.1-3. *If A is an accretive unbounded operator in a Hilbert space, then $\cos A = 0$.*

Proof. If $\operatorname{Re} \langle Ax, x \rangle$ is bounded above uniformly, $\cos A = 0$ immediately by the unboundedness of A; therefore, we may assume that there exists a sequence $\{u_n\}$, $\|u_n\| = 1$, $\operatorname{Re} \langle Au_n, u_n \rangle \to \infty$. Let $w_n = \eta_n u_n + \xi_n v_n$, where $\eta_n = [\operatorname{Re} \langle Au_n, u_n \rangle]^{-\alpha}$, $\xi_n = (1 - \eta_n^2)^{1/2}$, $\frac{1}{2} \leq \alpha < 1$, and $v_n \in D(A)$, $\|v_n\| = 1$; v_n will be specifically chosen later. Then for all sufficiently large n, if $\|Av_n\|$ is uniformly bounded, one has by the inverse triangle inequality:

(3.1-5)
$$R(w_n) \equiv \frac{\operatorname{Re}\langle Aw_n, w_n \rangle}{\|Aw_n\| \cdot \|w_n\|}$$

$$\leq \frac{\xi_n^2 \operatorname{Re}\langle Av_n, v_n \rangle + \eta_n \xi_n \operatorname{Re}\langle Av_n, u_n \rangle + \eta_n \xi_n \operatorname{Re}\langle Au_n, v_n \rangle + \eta_n^2 \operatorname{Re}\langle Au_n, u_n \rangle}{[\eta_n \|Au_n\| - \xi_n \|Av_n\|] \cdot |\xi_n - \eta_n|}$$

$$= \frac{(N_1 + N_2 + N_3 + N_4)}{D},$$

with the denominator $D \to \infty$, since $|\xi_n - \eta_n| \to 1$, and $\|Au_n\| \cdot [\operatorname{Re}\langle Au_n, u_n \rangle]^{-\alpha} \to \infty$ for $0 < \alpha < 1$; the latter may be seen as follows. Let $\|u_n\| = 1$, $\alpha < 1$, and $\operatorname{Re}\langle Au_n, u_n \rangle \to \infty$. Then, by Schwarz's inequality, one has $\|A \cdot u_n\|^2 \cdot [\operatorname{Re}\langle Au_n u_n \rangle]^{-2\alpha} \geq [\operatorname{Re}\langle Au_n u_n \rangle]^{2-2\alpha} \to \infty$.

Let us now consider the four terms N_i/D separately. If $\|Av_n\|$ is uniformly bounded, clearly (by Schwarz's inequality) $N_1/D \to 0$ and $N_2 \to 0$.

Also, $N_4 = 1$ if $\alpha = 1/2$, $N_4 \to 0$ if $\alpha > 1/2$; thus $(N_1 + N_2 + N_4) \cdot D^{-1} \to 0$ for $1/2 \leq \alpha < 1$. Therefore, if $|N_3|$ is uniformly bounded, $R(w_n) \to 0$ in (3.1-5). Now, if there exists at least one nontrivial vector $v \in D(A) \cap D(A^*)$ (let it have norm $= 1$), then taking $v_n \equiv v$, $\|Av_n\|$ is obviously uniformly bounded, and $|N_3| = \eta_n \xi_n |\text{Re}\langle u_n, A^* v_n\rangle| \leq \|A^* v\|$. If $D(A) \cap D(A^*) = \{0\}$, we may proceed as follows. Select $x, y \in D(A)$, $\|x\| = \|y\| = 1$, $\langle x, y\rangle = 0$, and let $v_n = \alpha_n x + \beta_n y$, $|\alpha_n|^2 + |\beta_n|^2 = 1$. Now choose α_n, β_n so that $\langle v_n, Au_n\rangle = 0$; that this can always be done is assured by taking α_n and β_n from the solutions of the equation $\alpha_n \langle x, Au_n\rangle + \beta_n \langle y, Au_n\rangle = 0$. Then $\|v_n\| = 1$, $\|Av_n\| \leq \|Ax\| + \|Ay\|$, $N_3 = 0$, and $R(w_n) \to 0$. □

For bounded strongly *accretive* operators T (Re $\langle Tx, x\rangle \geq m_T > 0$ for $\|x\| = 1$), we know by Theorem 3.1-1 that the angle $\phi(T)$ is less than $90°$.

Example. Consider the 2×2 matrix

$$A = \begin{bmatrix} 1 & 0 \\ 0 & 2 \end{bmatrix}.$$

We have $\cos A = 2\sqrt{2}/3 \cong .94281$ and A thus has angle $\phi(A) \cong 19.471$ degrees.

Notes and References for Section 3.1

The angle of an operator was introduced in
K. Gustafson (1968a). "The Angle of an Operator and Positive Operator Products," *Bull. Amer. Math. Soc.* **74**, 488–492.
in connection with problems in the perturbation theory of semigroup generators, which were treated by
K. Gustafson (1968b). "Positive (Noncommuting) Operator Products and Semigroups, *Math. Zeit.* **105**, 160–172.
K. Gustafson (1968c). "A Note on Left Multiplication of Semigroup Generators, *Pacific J. Math.* **24**, 463–465.

The formula (1968a, Corollary 4.4) for $\cos T$ for T selfadjoint was obtained by the norm convexity techniques of Gustafson (1968b, Lemma 1.3). The connection to the Kantorovich inequality was not known at that time.

In that connection, let us take this opportunity to note a typographical error in Gustafson (1968b), which could otherwise lead to confusion when reading those earlier papers. The inequality string two-thirds down on p. 163 appears as

$$\|\epsilon B + I\| \leq \|I - \epsilon^2 B^2\| \cdot \|\epsilon B - I\|^{-1} \leq (1 + \epsilon^2 \|B\|^2) \cdot [1 - \epsilon \theta(B)]^{-1}.$$

It should read

$$\|\epsilon B + I\| \leq \|I - \epsilon^2 B^2\| \cdot \|(\epsilon B - I)^{-1}\| \leq (1 + \epsilon^2 \|B\|^2) \cdot [1 - \epsilon \theta(B)]^{-1}.$$

There B is a bounded, *dissipative* (Re $[Bx, x] \leq 0$) operator on a Banach space X, $\epsilon \geq 0$, and $\theta(B) = \sup \text{Re}\,[Bx, x]$, $\|x\| = 1$, is its upper bound.

This typographical error does not affect the validity of any of the results in that paper.

The fact that $\cos T = 0$ for T unbounded was shown by
K. Gustafson and B. Zwahlen (1969). "On the Cosine of Unbounded Operators," *Acta Sci. Math.* **30**, 33–34.
This was somewhat generalized by
P. Hess (1971). "A Remark on the Cosine of Linear Operators," *Acta Sci. Math.* **32**, 267–269.
An example of $\cos T \neq 0$ for T unbounded in a Banach space was given in
K. Gustafson and M. Seddighin (1989). "Antieigenvalue Bounds," *J. Math. Anal. Appl.* **143**, 327–340.

3.2 Minmax Equality

The condition that B be strongly accretive, $\text{Re}\langle Bx, x\rangle \geq m_B > 0$ for $\|x\| = 1$, is equivalent to the existence of an interval of $\epsilon > 0$ such that $\|\epsilon B - I\| < 1$. The minimum

$$(3.2\text{-}1) \qquad g_m(B) = \min_{\epsilon > 0} \|\epsilon B - I\|$$

was of interest in semigroup perturbation theory, which started this operator trigonometry. We will defer an account of that semigroup theory to Section 3.4. For that theory, but more importantly for the development of the operator trigonometry, the following minmax result is important.

Theorem 3.2-1 (Minmax equality). *For a strongly accretive, bounded operator B on a Hilbert space,*

$$(3.2\text{-}2) \qquad \sup_{\|x\| \leq 1} \inf_{\epsilon} \|(\epsilon B - I)x\|^2 = \inf_{\epsilon > 0} \sup_{\|x\| \leq 1} \|(\epsilon B - I)x\|^2.$$

In particular,

$$(3.2\text{-}3) \qquad \sin B = g_m(B),$$

where the sine of B is defined by $\sin B = \sqrt{1 - \cos^2 B}$.

Proof. Let us first note that the right-hand side of (3.2-2) is just $g_m^2(B)$. We will show later that this minimum is attained uniquely. Also, we note that the left-hand side of (3.2-2) is indeed $1 - \cos^2 B$. To see this, consider the parabola

$$(3.2\text{-}4) \qquad \|(\epsilon B - I)x\|^2 = \epsilon^2 \|Bx\|^2 - 2\epsilon \text{Re}\langle Bx, x\rangle + 1.$$

This parabola achieves its minimum at the value $\epsilon_m(x) = \text{Re}\langle Bx, x\rangle / \|Bx\|^2$, and the value of the minimum is

$$(3.2\text{-}5) \qquad 1 - (\text{Re}\langle Bx, x\rangle / \|Bx\|)^2.$$

54   3. Operator Trigonometry

The supremum over x, $\|x\| = 1$, of this quantity is $1 - \cos^2 B$.

Next, let us assure ourselves that the minimum $g_m(B)$ is attained uniquely. This is the case for uniformly convex Banach spaces but not generally true otherwise. Here we give a Hilbert space proof. Let us suppose the contrary, that $\|\epsilon B - I\|$ dips down and then has a flat interval minimum. Let $\epsilon_1 < \epsilon_2$ be the ends of this interval. Then, for every small $\delta > 0$ there is an $x, \|x\| = 1$, such that the parabola $\|(\epsilon B - I)x\|^2$ is less than or equal to $g_m^2(B)$ at both ϵ_1 and ϵ_2 but at their midpoint must be within δ of the minimum flat:

(3.2-6) $$\|((\epsilon_1 + \epsilon_2)/2)B - I)x\|^2 \geq g_m^2(B) - \delta.$$

But no quadratic function of ϵ, anchored at the value 1 at $\epsilon = 0$, can satisfy this condition for arbitrarily small δ.

Let ϵ_m denote the unique $\epsilon > 0$ at which $g_m(B)$ is attained. The convex curve $\|\epsilon B - I\|^2$ is continuous in ϵ, it has left- and right-hand derivatives for all ϵ, and these are equal except at a countable set of ϵ. Thus, we may speak freely of the "slope" of $\|\epsilon B - I\|^2$ for a dense set of $\epsilon > 0$.

Consider now the curve $\|\epsilon B - I\|^2$ near its unique minimum value $g_m^2(B)$. For any chosen fixed ϵ just, but strictly, to the left of ϵ_m, $\|\epsilon B - I\|^2$ slopes downward and is strictly greater than $\|\epsilon_m B - I\|^2$. Since $\|\epsilon B - I\|^2$ is a supremum, there thus exists an $x_1, \|x_1\| = 1$, such that $\|(\epsilon B - I)x_1\|^2 > \|\epsilon_m B - I\|^2$. Moreover, we may choose this x_1 so that its minimum $\epsilon_m(x_1)$ lies (possibly nonstrictly) to the right of ϵ_m, for otherwise all of the parabolas $\|(\epsilon B - I)x_1\|^2$ increasing to achieve the supremum $\|(\epsilon B - I)\|^2$ from below would turn upward from points $\epsilon_m(x_1)$ strictly to the left of ϵ_m, cutting and thereby violating the downward-sloping curve $\|\epsilon B - I\|^2$. Similarly, there exists an $x_2, \|x_2\| = 1$, for any chosen fixed ϵ just, but strictly, to the right of ϵ_m, such that $\|(\epsilon B - I)x_2\|^2 > \|\epsilon_m B - I\|^2$ and such that $\epsilon_m(x_2)$ lies (possibly nonstrictly) to the left of ϵ_m. Moreover, we can choose x_1 and x_2 so that $\|(\epsilon B - I)x_1\|^2 \geq g_m^2(B) - \delta$ and $\|(\epsilon_m B - I)x_2\|^2 \geq g_m^2(B) - \delta$ for any prespecified small $\delta > 0$, so that those two parabolas pass arbitrarily close below the minimum point $g_m^2(B)$.

Consider first the case in which, for given small $\delta > 0$, $\epsilon_m(x_1)$ and $\epsilon_m(x_2)$ lie strictly to the right and left of ϵ_m, respectively. Let $x = \xi x_1 + \eta x_2$, where ξ and η are real and satisfy

(3.2-7) $$1 = \|x\|^2 = \xi^2 + \eta^2 + 2\eta\xi \operatorname{Re} \langle x_1, x_2 \rangle.$$

Using (3.2-7), a simple computation shows that $\|(\epsilon_m B - I)x\|^2 \geq g_m^2(B) - \delta + 2\xi\eta C$, where $C = \operatorname{Re}\{\langle(\epsilon_m B - I)x_1, (\epsilon_m B - I)x_2\rangle - (g_m^2(B) - \delta)\langle x_1, x_2\rangle\}$, and by restricting ξ and η in (3.2-7) to the appropriate quadrant, we can assure that $2\xi\eta C \geq 0$. By choosing ξ and η not only of appropriate sign but also near one or the other coordinate axes, we can also assure at this point in the construction that this term $2\xi\eta C$ is also arbitrarily small, but it turns out better to let this happen automatically as a consequence of later steps in the construction.

Next, we would like to show that $\epsilon_m(x)$ can be made arbitrarily close to ϵ_m. Consider first the case in which we ask for exact equality, $\epsilon_m(x) = \epsilon_m$. By a short computation, this can be seen to be equivalent to

$$\begin{aligned}(3.2\text{-}8)\quad &\xi^2 \left\{ \operatorname{Re} \langle Bx_1, x_1 \rangle \left(1 - \frac{\epsilon_m}{\epsilon_1}\right) \right\} + \eta^2 \left\{ \operatorname{Re} \langle Bx_2, x_2 \rangle \left(1 - \frac{\epsilon_m}{\epsilon_2}\right) \right\} \\ &- 2\xi\eta \operatorname{Re} \langle Bx_1, (\epsilon_m B - I)x_2 \rangle = 0,\end{aligned}$$

where ϵ_1 denotes $\epsilon_m(x_1)$ and ϵ_2 denotes $\epsilon_m(x_2)$. Since $(1 - \epsilon_m/\epsilon_1)(1 - \epsilon_m/\epsilon_2) < 0$, the degenerate hyperbola (3.2-8) and the ellipse (3.2-7) have a point in common. This assures that $\epsilon_m(x) = \epsilon_m$ and that $\|(\epsilon_m B - I)x\|^2 \geq g_m^2(B) - \delta$ for arbitrarily small $\delta > 0$. But since $\epsilon_m(x) = \epsilon_m$, the term $2\xi\eta C$ must be small, for otherwise $\|(\epsilon_m B - I)x\|^2$ would exceed $\|\epsilon_m B - I\|^2$, which it cannot. Thus these $x = x(\delta)$ provide the supremum sequence for the left side of (3.2-2) to equal the right side, namely, $g_m^2(B)$.

For those cases in which we cannot ask for $\epsilon_m(x)$ to be exactly ϵ_m in the above construction, but only arbitrarily close to ϵ_m, we proceed in the same way. It is desired that

$$(3.2\text{-}9)\quad \epsilon_m(x) = \frac{\xi^2 \operatorname{Re} \langle Bx_1, x_1 \rangle + \eta^2 \operatorname{Re} \langle Bx_2, x_2 \rangle + 2\xi\eta \operatorname{Re} \langle Bx_1, x_2 \rangle}{\xi^2 \|Bx_1\|^2 + \eta^2 \|Bx_2\|^2 + 2\xi\eta \operatorname{Re} \langle Bx_1, Bx_2 \rangle} = \tilde{\epsilon}_m,$$

where $\tilde{\epsilon}_m$ is to be arbitrarily close to ϵ_m. Multiplying this out results in the same expression (3.2-8), with ϵ_m replaced by $\tilde{\epsilon}_m$. Provided that we are not in an exceptional instance in which $\epsilon_1 = \epsilon_2$, we may ask for $\tilde{\epsilon}_m$ arbitrarily close to ϵ_m strictly between ϵ_2 and ϵ_1 and proceed as before.

The special cases in which $\epsilon_m(x_1)$ or $\epsilon_m(x_2)$ coincides with ϵ_m correspond to one or both branches of (3.2-8) lying along ξ or η coordinate axes. In that case, one chooses $\xi = 0$ or $\eta = 0$ as the case may be; it happens then that the $2\xi\eta C$ term is killed exactly. The special case in which $x_1 = x_2$ corresponds to a degenerate parabola in (3.2-7). These two parallel lines span all four quadrants, so the sign choice on ξ, η remains unrestricted. Thus, in all cases we have constructed x with $\epsilon_m(x)$ arbitrarily close to ϵ_m and $\|(\epsilon_m B - I)x\|^2$ arbitrarily close below $g_m^2(B)$. The supremum of such x shows that the left side of (3.2-2) attains the right side. □

Example. Consider the 2×2 matrix

$$A = \begin{bmatrix} 1 & 0 \\ 0 & 2 \end{bmatrix}$$

considered previously. For positive, selfadjoint operators, one always has

$$\|\epsilon A - I\| = \max\{1 - \epsilon m, \epsilon M - 1\}.$$

These two lines intersect at $\epsilon_m = 2/3$ to provide $g_m(A) = \frac{1}{3}$. Thus, according to Theorem 3.2-1, one has $\sin(A) = \frac{1}{3}$. On the other hand, from Theorem 3.1-2, we have $\cos(A) = 2\sqrt{2}/3$. The compatibility of these two characterizations is seen in $\sin^2 A + \cos^2 A = 1$.

56   3. Operator Trigonometry

Notes and References for Section 3.2

The minmax result was given in
K. Gustafson (1968d). "A Min-Max Theorem," *Notices Amer. Math. Soc.* **15**, 799.
  The proof given here is that of
K. Gustafson (1995). "Matrix Trigonometry," *Linear Algebra Appl.* **217**, 117–140.
In the related paper
E. Asplund and V. Ptak (1971). "A Minimax Inequality for Operators and a Related Numerical Range," *Acta Math.* **126**, 53–62.
it was shown that a generalization of the min-max result to $\epsilon A + B$ holds, but only in a Hilbert space.

3.3 Operator Deviations

Krein (1969; see Notes) introduced, in a different semigroup context, a quantity he called the *deviation* of T, dev(T). This quantity is equivalent to $\phi(T)$. There the following lemma was stated without proof.

Lemma 3.3-1. *Let x, y, z be three unit vectors in a Hilbert space. Define the angles $\phi_{xy}, \phi_{yz}, \phi_{xz}$ by $\cos\phi_{xy} = \text{Re}\langle x, y\rangle$, $\cos\phi_{yz} = \text{Re}\langle y, z\rangle$, $\cos\phi_{xz} = \text{Re}\langle x, z\rangle$, respectively, with $0 \leqq \phi_{xy}, \phi_{yz}, \phi_{xz} \leqq \pi$. Then*

$$(3.3\text{-}1) \qquad \phi_{xz} \leqq \phi_{xy} + \phi_{yz}.$$

Proof. Let $\langle x, y\rangle = a_1 + ib_1$, $\langle y, z\rangle = a_2 + ib_2$, $\langle x, z\rangle = a_3 + ib_3$, where the a_i, $i = 1, 2, 3$, and b_i, $i = 1, 2, 3$, are real and $|a_i|^2 + |b_i|^2 \leq 1$ for $i = 1, 2, 3$. We have $\cos\phi_{xy} = a_1$, $\cos\phi_{yz} = a_2$, and $\cos\phi_{xz} = a_3$. Since $\cos\alpha$ is a decreasing function of α in the interval $0 \leq \alpha \leq \pi$, we need to prove only that

$$\cos\phi_{xz} \geq \cos(\phi_{xy} + \phi_{yz})$$

or equivalently,

$$a_3 \geq a_1 a_2 - \sqrt{1 - a_1^2}\sqrt{1 - a_2^2},$$

where $\sin\phi_{xy} = \sqrt{1 - a_1^2} \geq 0$ and $\sin\phi_{yz} = \sqrt{1 - a_2^2} \geq 0$. Thus, we need

$$(3.3\text{-}2) \qquad \sqrt{1 - a_1^2}\sqrt{1 - a_2^2} \geq a_1 a_2 - a_3.$$

The result is obvious if the expression on the right of (3.3-2) is negative. If it is nonnegative, we need to prove that

$$(1 - a_1^2)(1 - a_2^2) \geq (a_1 a_2 - a_3)^2$$

or

(3.3-3) $$1 - a_1^2 - a_2^2 - a_3^2 + 2a_1 a_2 a_3 \geq 0.$$

Since the matrix

(3.3-4) $$G = \begin{bmatrix} \langle x,x \rangle & \langle x,y \rangle & \langle x,z \rangle \\ \langle y,x \rangle & \langle y,y \rangle & \langle y,z \rangle \\ \langle z,x \rangle & \langle z,y \rangle & \langle z,z \rangle \end{bmatrix}$$

is positive semidefinite, as is its entrywise complex conjugate, the matrix

(3.3-5) $$\begin{bmatrix} 1 & a_1 & a_3 \\ a_1 & 1 & a_2 \\ a_3 & a_2 & 1 \end{bmatrix}$$

is positive semidefinite, so its determinant is nonnegative, yielding (3.3-3). □

One now defines

$$\operatorname{dev}(T) = \sup_{x \in H} \phi(Tx, x),$$

where $\phi(Tx, x)$, $0 \leq \phi \leq \pi$, is defined by the equation

$$\cos(\phi(Tx, x)) = \frac{\operatorname{Re} \langle Tx, x \rangle}{\|Tx\| \cdot \|x\|}.$$

Notice that $\operatorname{dev}(T)$ is thus the same as the operator angle $\phi(T)$ of Section 3.1. From the triangle inequality of Lemma 3.3-1, one then obtains the following.

Theorem 3.3-2. *Let A and B be bounded invertible operators in a Hilbert space. Then*

(3.3-6) $$\phi(AB) \leq \phi(A) + \phi(B).$$

Proof. From Lemma 3.3-1, we have

$$\phi(ABx, x) \leq \phi(ABx, A^{-1}x) + \phi(A^{-1}x, x)$$
$$= \phi(Bx, x) + \phi(Ax, x).$$

Their suprema bear the same relation, so one has

$$\operatorname{dev}(AB) \leq \operatorname{dev}(B) + \operatorname{dev}(A). \quad \square$$

This fact, that the maximum turning angle of AB is less than or equal to that of A followed by that of B, is of course geometrically evident.

Notes and References for Section 3.3

Lemma 3.3-1 and Theorem 3.3-2 are due to

M. Krein (1969). "Angular Localization of the Spectrum of a Multiplicative Integral in a Hilbert Space," *Functional Anal. Appl.* **3**, 89–90.

The proof of Lemma 3.3-1 was apparently first given by

D. Rao (1972). *Numerical Range and Positivity of Operator Products*, Ph.D. thesis, University of Colorado, Boulder, Colorado.

See also

D. Rao (1976). "A Triangle Inequality for Angles in a Hilbert Space," *Revista Colombiana de Matematicas* **10**, 95–97.

In the proof of Lemma 3.3-1 given here, we have incorporated a simplification suggested by T. Ando (1993, private communication).

There is a third entity, independently introduced by H. Wielandt, at about the same time, which is also essentially equivalent to the angle $\phi(T)$ and the deviation dev (T). In his lecture notes,

H. Wielandt (1967). *Topics in the Analytic theory of Matrices*, University of Wisconsin Lecture Notes, Madison, Wisconsin.

the maximum angle

$$\gamma(A) = \sup \angle(x, Ax), \qquad Ax \neq 0$$

between vectors x and Ax is called the *singular angle* of a square matrix. Wielandt also used (omitting proof, as Krein did) a triangle inequality equivalent to (3.3-1). Wielandt's motivation, unlike that of Gustafson or Krein, both of whom were working on different (see the next section) semigroup problems, seems to have been in seeking generalizations of the Weyl theory for spectra of $\sigma(A + B)$ in terms of $\sigma(A)$ and $\sigma(B)$.

Although the operator angle $\phi(T)$ was the first to appear in the refereed journals, it seems almost certain that the three ideas $\phi(T)$, dev (T), and $\gamma(T)$ were all conceived independently, in different contexts, in 1967. A full accounting of this may be found in

K. Gustafson (1996). "Operator Angles (Gustafson), Matrix Singular Angles (Wielandt), Operator Deviations (Krein)," *Collected Works of Helmut Wielandt, II* (B. Huppert and H. Schneider, eds.), De Gruyters, Berlin.

3.4 Semigroup Generators

It is interesting that both the angle $\phi(T)$ and the deviation dev (T) came from the incubator of semigroup theory. The circumstances were, however, completely different. Let us briefly describe the motivations in the semigroup theory that led to $\phi(T)$ and dev (T). Recall that it is desired, for general unbounded operators T (e.g., differential operators), to define the semigroup by

$$(3.4\text{-}1) \qquad e^{tT} = s\text{-}\lim_{n \to \infty} \left(I - \frac{t}{n} T\right)^{-n}.$$

Conversely, given a semigroup $U(t)$, it is desired to characterize its infinitesimal generator T by

$$(3.4\text{-}2) \qquad Tx = s\text{-}\lim_{t\to 0} \frac{(U(t)-I)x}{t}.$$

We refer to the references at the end of this section for further details of the semigroup theory, as we will use them in the following discussion.

In particular, let us recall the characterization theorem for the contraction semigroup generators $G(1,0)$ due to Hille and Yosida as modified by Phillips and Lumer, namely, *Theorem (HYPL)*: $A \in G(1,0)$ iff A is dissipative and $\lambda I - A$ is onto for all λ with $\operatorname{Re}\lambda > 0$. This theorem holds generally for a Banach space X equipped with a semi-inner product $[x,y]$a, and an operator A is called dissipative if $\operatorname{Re}[Ax,x] \leqq 0$ for all $x \in D(A)$. This class $G(1,0)$ of generators of contraction semigroups $\|e^{tT}\| \leqq 1$ is the most important in applications, and other classes of higher-growth semigroups can often be reduced to it or treated by similar methods. The contraction property generalizes the unitarity of the semigroups e^{itA} in the case that $T = iA$, A a selfadjoint operator in a Hilbert space.

Thus, in Hilbert space a necessary condition for A to be such a semigroup generator is that the numerical range $W(A)$ be in the left half (closed) plane. Applications often require that one work with perturbations of simpler generators A, and one wants to know when, for example, $A+B$ remains a generator when A is perturbed by some relatively subordinate operator B. A somewhat final result in that direction is given by *Theorem (RKNG)*: if $A \in G(1,0)$ and B is a relatively small perturbation, $\|Bx\| \leqq a\|Ax\| + b\|x\|$ on $D(A)$, where $a < 1$, and $A + B \in G(1,0)$ iff $A + B$ is dissipative. The RKNG stands for Rellich–Kato–Nagy–Gustafson. A similar multiplicative perturbation result may be obtained directly from the RKNG theorem. The multiplicative perturbation question may be posed: given $A \in G(,0)$, when is $BA \in G(1,0)$? If one notes that $BA \in G(1,0)$ iff $\epsilon BA \in G(1,0)$ for arbitrary $\epsilon > 0$, then one can write the multiplicative perturbation as an additive perturbation $\epsilon BA = (\epsilon B - I)A + A$ and ask when $(\epsilon B - I)A$ satisfies the additive perturbation requirement of the RKNG theorem. Restricting attention to the case of B bounded and everywhere defined, we have the following theorem.

Theorem 3.4-1 (Multiplicative perturbation). *For X a Banach space with semi-inner product $[x,y]$ and $A \in G(1,0)$ an infinitesimal generator of a contraction semigroup on X, if there is an $\epsilon > 0$ such that $\|\epsilon B - I\| < 1$, then $BA \in G(1,0)$ iff BA is dissipative.*

Proof. As described above, write $\epsilon BA = (\epsilon B - I)A + A$ and employ the RKNG additive perturbation theorem. □

Turning then to the remaining question of when BA is dissipative, recall that the condition $\|\epsilon B - I\| < 1$ for some $\epsilon > 0$ is equivalent to B being strongly accretive: $\operatorname{Re} \langle Bx, x \rangle \geqq m_B > 0$ for $\|x\| = 1$. This condition is, of course, entirely independent of A. But if we write

$$\begin{aligned}(3.4\text{-}3)\qquad \operatorname{Re} \langle \epsilon B A x, x \rangle &= \operatorname{Re} \langle (\epsilon B - I) A x, x \rangle + \operatorname{Re} \langle A x, x \rangle \\ &\leqq \|\epsilon B - I\| \|A x\| \|x\| + \operatorname{Re} \langle A x, x \rangle,\end{aligned}$$

we see that BA will be dissipative if $\|\epsilon B - I\| \leqq \cos(-A)$. This began the operator trigonometry. The second key development of the operator trigonometry came about by understanding, through the minmax theorem, (Theorem 3.2-1), that $g_m(B) = \min \|\epsilon B - I\|$, $\epsilon > 0$, is $\sin B$. Thus a sufficient condition for BA to be dissipative when B is accretive and A is dissipative is

$$(3.4\text{-}4)\qquad\qquad \sin B \leqq \cos(-A).$$

We will elaborate on this in the next section.

Similarly, it is interesting to see how the quantity $\operatorname{dev}(T)$ arose out of a semigroup theory. The main need was to bound $\sigma(T)$ in a sector: $|\arg \lambda| \leqq \operatorname{dev}(T)$. This came about in attempting to integrate an initial-value problem

$$(3.4\text{-}5)\qquad \begin{cases} \dfrac{dx(t)}{dt} = A(t) x(t), \\ x(0) = x_0, \end{cases}$$

where $A(t)$ is a time-dependent, infinitesimal generator found as the derivative of a function $F(t)$ of strongly bounded variation. Then the integral $W(t) = \int_0^t \exp\{dF(t)\}$ solves the initial-value problem (3.4-5) with solution $x(t) = W(t) x_0$. Since $W(t)$ is in fact a multiplicative integral due to the multiplicative property of the exponentials, written formally, $W = \Pi_0^t \exp dF$, then by Theorem 3.3-2 above one has, again formally,

$$(3.4\text{-}6)\qquad\qquad \operatorname{dev}(W) \leqq \sum_0^t \operatorname{dev} \exp dF.$$

Since $\sigma(W)$ is in a sector bounded by $\operatorname{dev}(W)$, this guarantees, for example, that the negative real axis is not in $\sigma(W)$. Recall that this is a useful property when integrating systems of ordinary differential equations, and the abstract semigroup theory follows similarly.

Let us note that when dF happens to be a selfadjoint operator H, then

$$(3.4\text{-}7)\qquad\qquad \cos \operatorname{dev} \exp dF = \frac{2\sqrt{e^m e^M}}{e^m + e^M},$$

where m and M are the lower and upper bounds of H, respectively. Taking the largest and smallest M and m over the whole family of H in dF then provides a uniform bound for $\operatorname{dev}(W)$ in (3.4-6), and hence a uniform bound on $\sigma(W(t))$ for the solution of the initial-value problem (3.4-5).

Notes and References for Section 3.4

Good references for the semigroup theory are
T. Kato (1980). *Perturbation Theory for Linear Operators*, 2nd Ed., Springer, New York.
E. Hille and R. Phillips (1957). *Functional Analysis and Semigroups*, Amer. Math. Soc. Colloq. Publ. **31**, Providence, RI.
The Lumer–Phillips version of the Hille–Yosida theorem is given in
G. Lumer and R. Phillips (1961). "Dissipative Operators in a Banach Space," *Pacific J. Math.* **11**, 679–698.
For the theory of time dependent semigroup generators, see
K. Yosida (1966). *Functional Analysis*, Springer, Berlin.
  For additive perturbation results, see
K. Gustafson (1966). "A Perturbation Lemma," *Bull. Amer. Math. Soc.* **72**, 334–338.
K. Gustafson (1983). "The RKNG (Rellich–Kato–Nagy–Gustafson) Perturbation Theorem for Linear Operators in Hilbert and Banach Space," *Acta Sci. Math.* **45**, 201–211.
For multiplicative perturbation results, see
K. Gustafson (1968c). "A Note on Left Multiplication of Semigroup Generators," *Pacific J. Math.* **24**, 463–465.
K. Gustafson and Ken-iti Sato (1969). "Some Perturbation Theorems for Nonnegative Contraction Semigroups," *J. Math. Soc. Japan* **21**, 200–204.
K. Gustafson and G. Lumer (1972). "Multiplicative Perturbation of Semigroup Generators," *Pacific J. Math..* **41**, 731–742.
For recent results, see
R. Kumar and P. Das (1991). "Perturbation of m-Accretive Operators in Banach Spaces," *Nonlinear Analysis* **17**, 161–168.
  For the origin of dev(T) in the semigroup context, see
M. Krein (1969). "Angular Localization of the Spectrum of a Multiplicative Integral in a Hilbert Space," *Functional Anal. Appl.* **3**, 89–90.

3.5 Accretive Products

In the semigroup perturbation theorems of the preceding section, the two issues were that a perturbed generator $T = A + B$ or $T = BA$ remain maximal in the sense that $\lambda I - T$ is surjective for $\operatorname{Re} \lambda > 0$, and that T remain dissipative: $\operatorname{Re} \langle Tx, x \rangle \leqq 0$. The maximality follows from the now well-established index (Fredholm) theory of linear operators. The dissipativeness of the sum $A + B$ can often be seen quickly, due to the subadditivity of the numerical range

$$W(A+B) \subset W(A) + W(B)$$

62   3. Operator Trigonometry

But the same cannot be said of the product BA. Specifically, when $\operatorname{Re}\langle Ax, x\rangle \leqq 0$ and $\operatorname{Re}\langle Bx, x\rangle \geqq 0$, can one expect that $\operatorname{Re}\langle ABx, x\rangle \leqq 0$? As the example given at the beginning of Chapter 2 shows, this is generally not the case.

In this section, therefore, we consider the question of when the product (i.e., composition) of two accretive operators is itself accretive. It follows from the preceding section that the operator trigonometry provides an interesting sufficient condition. Indeed, this question constituted the inception of the operator trigonometry.

Lemma 3.5-1. *Let B and A be bounded, accretive operators in a Hilbert space. Then BA is accretive if*

$$(3.5\text{-}1) \qquad \sin B \leqq \cos A.$$

Proof. The inequality (3.4-3) and our understandings of $\sin B$ and $\cos A$. □

To get more feeling about the sharpness of this operator trigonometry as applied to the accretive operator product question, let us note how it improves what one might obtain otherwise. For example, if one writes for strongly accretive B and A and any x

$$\operatorname{Re}\langle BAx, x\rangle = \operatorname{Re}\langle (B - I)Ax, x\rangle + \operatorname{Re}\langle Ax, x\rangle,$$

then, taking $\|x\| = 1$ for convenience, bounding the two terms on the right-hand side in the natural way,

$$(3.5\text{-}2) \qquad \begin{aligned} |\operatorname{Re}\langle (B-I)Ax, x\rangle| &\leqq \|(B-I)\|\|A\| \quad \text{and} \\ \operatorname{Re}\langle Ax, x\rangle &\geqq m(A), \end{aligned}$$

leads to the sufficient condition for BA to be accretive by (a): $\|B - I\| \leqq m(A)/\|A\|$. Given (a), we may improve the criteria for product accretivity by (b): put ϵB into (3.5-2) with ϵ optimal so that $\|\epsilon B - I\| = \sin B$ is now inserted into (3.5-2). A further sharpening, given (b), may be had by replacing (a) with (c): divide out the $\|Ax\|$ in (3.5-2), e.g., as in (3.4-3), thereby effectively inserting $\cos A$ into the inequalities for product accretivity. Having considered this criteria-sharpening sequence (a), (b), (c), let us illustrate it with an example.

Example. Let A and B be positive selfadjoint operators with $\|A\| = \|B\| = 1$ and $m(A) = 1/2$, $m(B) > 0$. From the sufficient condition

$$\|B - I\| \leqq m(A)/\|A\|,$$

one can assure that BA is accretive for $m(B)$ down to $m(B) = 1/2$. From the sufficient condition

$$\sin B \leqq m(A)/\|A\|,$$

one can assure that BA is accretive for $m(B)$ down to $m(B) = 1/3$. From the sufficient condition

$$\sin B \leq \cos A,$$

one can assure that BA is accretive for $m(B)$ down to $m(B) = 0.0295$.

In the example, we used the facts that $\|B - I\| = 1 - m(B)$, $g_m(B) = \min_\epsilon \|\epsilon B - I\| = \sin B = (\|B\| - m(B))/(\|B\| + m(B))$, and $\cos A \cong 0.9428$, so that we need only

$$\frac{1 - m(B)}{1 + m(B)} \leq 0.9428,$$

from which it follows that BA will be positive whenever $m(B) \geq 0.0295$. This is more than a 10-fold improvement over the first two sufficient conditions and says that "most" positive, selfadjoint B when multiplied against the given A will retain the positivity of BA.

The sufficient condition $\sin B \leq \cos A$ for a positive numerical range $W(BA)$ may be generalized to a notion of supersets for the whole numerical range $W(BA)$. For a bounded operator T on a complex Hilbert space, let us recall the abbreviated notations

$$W(T) = \{(Tx, x), \|x\| = 1\},$$
$$m_T = \inf_{\|x\|=1} |(Tx, x)|,$$
$$w_T = \sup_{\|x\|=1} |(Tx, x)|,$$
$$\theta_T = \sup_{z \in W(T)} |\arg z|,$$
$$\cos T = \inf_{Tx \neq 0} \operatorname{Re}(Tx, x)/\|x\| \cdot \|Tx\|,$$
$$|\cos|T = \inf_{Tx \neq 0} |(Tx, x)|/\|x\| \cdot \|Tx\|,$$
$$|\sin|T = [1 - |\cos|^2 T]^{1/2}.$$

Let D be the subset of the complex plane consisting of the semiannulus $z = re^{i\theta}$ with $m_A m_B \leq r \leq w_A w_B$, $|\theta| \leq \theta_A + \theta_B$.

Theorem 3.5-2. Let A and B be bounded operators on a Hilbert space. Then

(3.5-3)
$$W(AB) \subset \Sigma \bigcup \{0\}, \quad \text{where}$$
$$\Sigma = \bigcup_{\lambda \in D} \{z \mid |z - \lambda| \leq \|AB\|(|\sin|A)(|\sin|B)\}.$$

Proof. Let G denote the Grammian or Gram determinant $G = G(x_1, x_2, x_3)$ of three vectors in a Hilbert space, namely

$$G = \begin{vmatrix} \langle x_1, x_1 \rangle & \langle x_1, x_2 \rangle & \langle x_1, x_3 \rangle \\ \langle x_2, x_1 \rangle & \langle x_2, x_2 \rangle & \langle x_2, x_3 \rangle \\ \langle x_3, x_1 \rangle & \langle x_3, x_2 \rangle & \langle x_3, x_3 \rangle \end{vmatrix}.$$

G is always real and positive, and strictly positive iff the three vectors are linearly independent. Let $x_1 = ABx/\|ABx\|$, $x_2 = Bx/\|Bx\|$, and $x_3 = x$, a unit vector; we assume $ABx \neq 0$ here. Also, let $\langle x_1, x_2 \rangle = ae^{i\alpha}$, $\langle x_2, x_3 \rangle = be^{i\beta}$, and $\langle x_1, x_3 \rangle = \rho e^{i\gamma}$. Then by the positivity of G, we have

$$1 + 2\rho ab \cos(-\gamma + \alpha + \beta) - a^2 - b^2 - \rho^2 \geq 0$$

and thus

$$[\rho \cos\gamma - ab\cos(\alpha+\beta)]^2 + [\rho\sin\gamma - ab\sin(\alpha+\beta)]^2 \leq (1-a^2)(1-b^2).$$

Recalling that $\langle ABx, x \rangle = \|ABx\|\rho(\cos\gamma + i\sin\gamma)$, and because $a \geq |\cos|A$ and $b \geq |\cos|B$, we have

(3.5-4)
$$[\text{Re}\,\langle ABx, x\rangle - \text{Re}\,\lambda]^2 + [\text{Im}\,\langle ABx, x\rangle - \text{Im}\,\lambda]^2$$
$$\leq \|ABx\|^2 (|\sin|A)^2 (|\sin|B)^2,$$

where λ denotes the point $\|ABx\|ab(\cos(\alpha+\beta) + i\sin(\alpha+\beta))$. Since λ is in D, we have shown that $\langle ABx, x\rangle$ is in Σ.

Thus $W(AB) \subset \Sigma \cup \{0\}$. Moreover, if $\{0\}$ is attained (e.g., in the case where AB is singular), then by the convexity of $W(AB)$ and the above argument all rays from $\{0\}$ to other points in $W(AB)$ must be in Σ, which, being a closed set, must therefore contain $\{0\}$. Then we may conclude that $\overline{W(AB)} \subset \Sigma$. □

As we have mentioned previously, the general numerical range mapping theorems of Chapter 2 do not encompass the question of when T accretive implies T^2 accretive. Lemma 3.5-1 gives us the sufficient criteria that the *real* angle of T less than or equal to 45°. But note carefully that this is not the same as specifying that $W(T)$ be in the ±45° sector of the complex plane, and the counterexample given earlier shows that condition to be insufficient. The criteria of Theorem 3.5-2, although general, are seen by some sectorial examples to allow too large a containment superset in that situation. Therefore, we turn to the theory of sectorial forms to consider this question. Because that theory embraces both bounded and unbounded operators T, we are able to treat both in the following.

Let T be a densely defined *sectorial operator* in a complex Hilbert space with semiangle $\theta_T < \pi/2$ and vertex $\gamma_T = 0$. We refer the reader to (Kato, 1980; see the Notes) for further details that we employ concerning sectorial operators, and throughout we assume for simplicity that the vertex $\overset{\circ}{\gamma}_T = 0$. Let $\overset{\circ}{t}[u,v]$ be the sesquilinear form $\langle Tu, v\rangle$ with $D(\overset{\circ}{t}) = D(T)$ and with closure $t[u,v]$. $W(T)$ is dense in $W(t)$, the adjoint form t^* is also sectorial, and the two closed symmetric forms $\frac{1}{2}(t+t^*)$ and $1/2i)(t-t^*)$ are uniquely represented by selfadjoint operators $\operatorname{Re} T$ and $\operatorname{Im} T$ such that $\frac{1}{2}(t+t^*)[u,v] = \langle(\operatorname{Re} T)u,v\rangle$ for $u \in D(\operatorname{Re} T)$, $v \in D(t) = D(t^*) = D(t+t^*) = D(t-t^*)$, and $\frac{1}{2i}(t-t^*)[u,v] = \langle(\operatorname{Im} T)u,v\rangle$ for $u \in D(\operatorname{Im} T)$, $v \in D(t)$.

Theorem 3.5-3. *Let T be a sectorial operator in a Hilbert space such that $D(T) \subset D(\operatorname{Re} T) \cap D(\operatorname{Im} T)$ and for which there exists $b \leq 1$ such that $\|(\operatorname{Im} T)u\| \leq b\|(\operatorname{Re} T)u\|$ for all $u \in D(T^2)$. Then T^2 is accretive and $\theta_{T^2} \leq 2\tan^{-1} b$.*

Proof. For $u \in D(T)$ and arbitrary $v \in D(T)$, we have

$$\langle Tv, u\rangle = \frac{1}{2}(t+t^*)[v,u] + \frac{1}{2}(t-t^*)[v,u]$$
(3.5-5)
$$= \langle v, (\operatorname{Re} T - i\operatorname{Im} T)u\rangle,$$

so that $D(T) \subset D(T^*)$ and $T^*u = (\operatorname{Re} T - i\operatorname{Im} T)u$ on $D(T)$. In particular, for $u \in D(T^2)$ we have

(3.5-6)
$$\langle T^2 u, v\rangle = \langle Tu, T^*u\rangle = \langle(\operatorname{Re} T + i\operatorname{Im} T)u, (\operatorname{Re} T - i\operatorname{Im} T)u\rangle$$
$$= \|(\operatorname{Re} T)u\|^2 - \|(\operatorname{Im} T)u\|^2 - 2i\langle(\operatorname{Re} T)u, (\operatorname{Im} u)\rangle.$$

Since $b \leq 1$, we have $\operatorname{Re}\langle T^2 u, r\rangle \geq 0$.

Concerning the sector angle for the numerical range of T^2, the case $b = 1$ is just a restatement of the accretivity of T^2, so any improvement comes only in the case $b < 1$. Also, we may assume $(\operatorname{Re} T)u \neq 0$, for otherwise $\langle T^2 u, u\rangle = 0$. Thus

(3.5-7)
$$\frac{|\operatorname{Im}\langle T^2 u, u\rangle|}{\operatorname{Re}\langle T^2 u, u\rangle} \leq \frac{2|\langle(\operatorname{Re} T)u, (\operatorname{Im} T)u\rangle|}{(1-b^2)\|(\operatorname{Re} T)u\|^2} \leq \frac{2b}{1-b^2} = \tan(2\tan^{-1} b),$$

which gives an improved bound on sector angle when $b < 1$. □

For a bounded operator, the above arguments yield the following necessary and sufficient condition for the accretivity of T^2. Note that in Theorem 3.5-3 and Corollary 3.5-4 we do not need to assume that T is accretive.

Corollary 3.5-4. *Let T be a bounded operator on a Hilbert space. Then T^2 is accretive iff $\|(\operatorname{Im} T)x\| \leq b\|(\operatorname{Re} T)x\|$ for some $b \leq 1$. In that case $\theta_{T^2} \leq 2\tan^{-1} b$.*

Proof. Immediate from Theorem 3.5-3. □

If in Theorem 3.5-3 T is m-sectorial, then T is its own Friedrich's extension. But it is not generally known when T can be written as $T = A + iB$, with $D(A) = D(B) = D(T)$, A and B symmetric. Via a partial result of this type we are able to obtain necessary and sufficient conditions for T^2 accretive for a certain class of m-sectorial operators T.

Theorem 3.5-5. *Let T be an m-sectorial operator with $D(T) \subset D((\operatorname{Re} T))$. Then $T \subset \operatorname{Re} T + iC$, where C is symmetric, and T^2 is accretive iff $\|Cx\| \leq b\|(\operatorname{Re} T)x\|$ for some $b \leq 1$ and all $x \in D(T^2)$. In that case, $\theta_{T^2} \leq 2\tan^{-1} b$.*

Proof. By the "angle-boundedness" factorization (Kato, 1980, p. 337), we may write $T = (\operatorname{Re} T)^{1/2}(I + IB)(\operatorname{Re} T)^{1/2}$, where B is selfadjoint and $\|B\| \leq \tan \theta_T$. By hypothesis, $x \in D(T)$ implies $(\operatorname{Re} T)^{1/2}x \in D((\operatorname{Re} T)^{1/2})$. Hence, by the construction of B, it may be verified that we have $x \in D(C)$, where $C = (\operatorname{Re} T)^{1/2} B (\operatorname{Re} T)^{1/2}$.

Since $T \subset \operatorname{Re} T + iC$, we have $T^* \supset \operatorname{Re} T - iC$, and for $x \in D(T^2)$ as in the proof of Theorem 3.5-3, we have

$$(T^2 x, x) = (Tx, (\operatorname{Re} T - iC)x)$$
$$= \|(\operatorname{Re} T)x\|^2 - \|Cx\|^2 - 2i\operatorname{Re}((\operatorname{Re} T)x, Cx),$$

the result following as before. □

Notes and References for Section 3.5

The sufficient condition $\sin B \leqq \cos A$ for operator product accretivity was first stated in
K. Gustafson (1968c). "A Note on Left Multiplication of Semigroup Generators," *Pacific J. Math.* **24**, 463–465.

The result of Theorem 3.3-2, see
M. Krein (1969). "Angular Localization of the Spectrum of a Multiplicative Integral in a Hilbert Space," *Functional Anal. Applic.* **3**, 89–90,
that $\operatorname{dev}(BA) \leqq \operatorname{dev}(B) + \operatorname{dev}(A)$, implies that BA is accretive when $\operatorname{dev}(B) + \operatorname{dev}(A) \leq \pi/2$, a hypothesis which implies the sufficient condition $\sin A \leq \cos B$ or $\sin B \leq \cos A$ of Lemma 3.5-1.

As mentioned in Section 2.5, the commuting case is much easier. For example, for positive selfadjoint A and bounded accretive B, BA is always accretive, as observed by
K. Gustafson (1968a). "The Angle of an Operator and Positive Operator Products," *Bull. Amer. Math. Soc.* **74**, 488–492.
Later this was extended to $W(AB) \subset W(A)W(B)$, without B necessarily being accretive but with A required bounded, by
R. Bouldin (1970). "The Numerical Range of a Product," *J. Math. Anal. Appl.* **32**, 459–467.

For A and B selfadjoint matrices, the positivity of A and that of $\operatorname{Re} AB$ necessitates that of B. Furthermore, it should be noted specifically that even when A and B are positive definite, selfadjoint operators, AB need not be accretive in the noncommuting case. For example, consider

$$A = \begin{bmatrix} 1+\alpha & 1-\alpha \\ 1-\alpha & 1+\alpha \end{bmatrix}, \quad B = \begin{bmatrix} 1+2\alpha & 2^{1/2}(\alpha-1) \\ 2^{1/2}(\alpha-1) & 2+\alpha \end{bmatrix}$$

where $\alpha > 0$. Then, for $\alpha = 2^{-1}(2^{12}-1)(5+2^{\frac{1}{2}})$ and $x = (1,0)$ one has

$$\langle ABx, x \rangle < 2^{-1}(1-2^{1/2}) < 0.$$

The superset Theorem 3.5-2 was obtained by

D. Rao (1972). *Numerical Range and Positivity of Operator Products*, Dissertation, University of Colorado, Boulder, Colorado.

For results on sectorial and sesquilinear forms, which we used in the proofs of Theorem 3.5-3 and Theorem 3.5-5, see

T. Kato (1980). *Perturbation Theory in Linear Operators*, 2nd Edition, Springer, New York.

A number of sufficient conditions beyond those presented here for positive operator products are given in

K. Gustafson and D. Rao (1977). "Numerical Range and Accretivity of Operator Products," *J. Math. Anal. Appl.* **60**, 693–702.

See also

K. Gustafson (1994). "Operator Trigonometry," *Linear and Multilinear Algebra* **37**, 139–159.

3.6 Antieigenvalue Theory

The quantity $\cos A$ has another interpretation as the first *antieigenvalue* of A:

$$(3.6\text{-}1) \qquad \mu_1(A) = \inf_{\substack{x \in D(A) \\ x \neq 0, \, Ax \neq 0}} \frac{\operatorname{Re}\langle Ax, x \rangle}{\|Ax\| \cdot \|x\|}.$$

Also defined were the higher antieigenvalues

$$(3.6\text{-}2) \qquad \mu_n(A) = \inf_{\substack{x \in D(A) \\ x \perp \{x_1, \ldots, x_{n-1}\}}} \frac{\operatorname{Re}\langle Ax, x \rangle}{\|Ax\| \|x\|},$$

where the x_k were called the corresponding *antieigenvectors* of A. It was proposed (see the Notes) to consider this as a spectral theory analogous to the usual spectral theory of eigenvalues and eigenvectors. *Total antieigenvalues*, e.g., the first one would be

$$(3.6\text{-}3) \qquad |\mu_1|(A) = \inf_{\substack{x \in D_A \\ x \neq 0 \\ Ax \neq 0}} \frac{|\langle Ax, x \rangle|}{\|Ax\| \cdot \|x\|}.$$

68   3. Operator Trigonometry

were also defined. When one thinks of eigenvalues and their corresponding eigenvectors, those are the vectors for which A dilates but does not turn at all. The name "antieigenvalues" was chosen to connote the opposite: the critical turnings of A.

By Theorem 3.1-2, we know for T strongly positive, selfadjoint that

$$\mu_1(T) = \frac{2\sqrt{\lambda_{\min}\lambda_{\max}}}{\lambda_{\min} + \lambda_{\max}},$$

where $\lambda_{\min} = m(T) =$ the lower bound of the numerical range $W(T)$, and $\lambda_{\max} = M(T) =$ the upper bound of $W(T)$. Thus, it is evident that generally one might expect relations between the antieigenvalues of T and the eigenvalues of T, the strength of that relationship perhaps decreasing as one departs from normal-like operators. Exact expressions for all of the antieigenvalues and antieigenvectors for strongly accretive, finite-dimensional normal operators have been obtained.

Theorem 3.6-1. *Let T be a normal accretive operator on a finite dimensional Hilbert space H with eigenvalues*

$$\lambda_i = \beta_i + i\delta_i, \quad i = 1, \ldots, n.$$

Let

$$E = \{\beta_i/|\lambda_i| : 1 \le i \le n\}$$

and

(3.6-4)
$$F = \Big\{ 2\frac{\sqrt{(\beta_j - \beta_i)(\beta_i|\lambda_j|^2 + 2\beta_j|\lambda_i|^2}}{|\lambda_j|^2 - |\lambda_i|^2}$$

$$0 \le \frac{\beta_j|\lambda_j|^2 - 2\beta_i|\lambda_j|^2 + 2\beta_j|\lambda_i|^2}{(|\lambda_i|^2 - |\lambda_j|^2)(\beta_i - \beta_j)} \le 1,$$

$$1 \le i \le n,\ 1 \le j \le n,\ i \ne j \Big\}.$$

Then $\mu_1(T)$ is exactly equal to the smallest number in $E \cup F$. Furthermore, if T is diagonal and

$$\mu_1(T) = 2\frac{\sqrt{(\beta_j - \beta_i)(\beta_i|\lambda_j|^2 - \beta_j|\lambda_i|^2)}}{|\lambda_j|^2 - |\lambda_i|^2}$$

then $\mu_1(T) = \langle Tz, z\rangle/\|Tz\|$, for some z with

(3.6-5)
$$|z_i|^2 = \frac{\beta_j|\lambda_j|^2 - 2\beta_i|\lambda_j|^2 + \beta_j|\lambda_i|^2}{(|\lambda_i|^2 - |\lambda_j|^2)(\beta_i - \beta_j)},$$

$$|z_j|^2 = \frac{\beta_i|\lambda_i|^2 - 2\beta_j|\lambda_i|^2 + \beta_i|\lambda_j|^2}{(|\lambda_i|^2 - |\lambda_j|^2)(\beta_i - \beta_j)},$$

and $z_k = 0$ for $k \ne i$, $k \ne j$.

3.6 Antieigenvalue Theory

Proof. See (Gustafson and Seddighin (1989), see Notes). The proof is somewhat involved and uses the Lagrange multiplier method. Note that it presumes that one knows all of the eigenvalues of T. □

Theorem 3.6-1 has been generalized to find all total antieigenvalues, including the higher ones, and their corresponding antieigenvectors. We state and prove this result first for the first total antieigenvalue. The proof for the higher antieigenvalues follows by the same methods.

Theorem 3.6-2. *Let T be a normal operator on a finite-dimensional Hilbert space H with eigenvalues $\lambda_i = \beta_i + i\,\delta_i$, $i = 1, \ldots, n$. Then, the first total antieigenvalue is either 1 or the smallest number in the set of values*

$$(3.6\text{-}6) \qquad G = \left\{ \frac{\sqrt{(\beta_i|\lambda_j| + \beta_j|\lambda_i|)^2 + (\delta - i|\lambda_j||\delta_j|\lambda_i|)^2}}{(|\lambda_i| + |\lambda_j|)\sqrt{|\lambda_i||\lambda_j|}}, \quad i,j = 1, \ldots, n \right\}.$$

Moreover, if T is diagonal and $|\mu_1(T)| = 1$, then the first total antieigenvector is $z^{(1)} = (z_1, \ldots, z_n)$, with $|z_j| = 1$ for some j and all other $z_i = 0$. If $|\mu_1(T)|$ is one of the values in G, then the components of z satisfy $|z_i|^2 = |\lambda_j|(|\lambda_i| + |\lambda_j|)^{-1}$, $|z_j|^2 = |\lambda_i|(|\lambda_i| + |\lambda_j|)^{-1}$, and all other $z_k = 0$.

Proof. We assume that T has already been diagonalized, i.e., that $\{e_1, \ldots, e_n\}$ is a basis for H with respect to which the normal operator T is diagonal. The statements of Theorem 3.6-2 are with respect to that basis. The theorem thus expresses the fact that the antieigenvectors of T are generally expressible in terms of just two of the eigenvectors of T. This is a considerable conceptual clarification as to the nature of antieigenvectors, at least for normal T, and as will be seen it also leads to easier computations of the higher antieigenvalues $|\mu_k|$. Of course, one would have to diagonalize T, that is, find all its eigenvectors first.

Let $z = (z_1, \ldots, z_n)$, $z_i = x_1 + i\,y_i$, be any vector in H, expressed in the eigenbasis $\{e_1, \ldots, e_n\}$ for T. Then the problem at hand is to minimize the positive function

$$(3.6\text{-}7) \qquad f(z) = \frac{|\langle Tz, z \rangle|^2}{\|Tz\|^2 \|z\|^2} = \frac{\left(\sum_{i=1}^{n} \beta_i |z_i|^2\right)^2 + \left(\sum_{i=1}^{n} \delta_i |z_i|^2\right)^2}{\sum_{i=1}^{n} |\lambda_i|^2 |z_i|^2}$$

on the set $\sum_{i=1}^{n} |z_i|^2 = 1$. First, suppose f attains its minimum at a point z on this sphere with only one $|z_j| = 1$, all others vanishing. Then $|\mu_1|(T) = 1$. Next, suppose f attains its minimum at a point z on this unit sphere with exactly two nonvanishing components, z_i and z_j. Then,

$$f(z) = \frac{\beta_i^2|z_i|^4 + \beta_j^2|z_j|^4 + 2\beta_i\beta_j|z_i|^2|z_j|^2 + \delta_i^2|z_i|^4 + \delta_j^2|z_j|^4 + 2\delta_i\delta_j|z_i|^2|z_j|^2}{|\lambda_i|^2|z_i|^2 + |\lambda_j|^2|z_j|^2}.$$

70 3. Operator Trigonometry

But in $|z_i|^2 + |z_j|^2 = 1$ we may let $x = |z_i|^2$, and then

$$f(z) = \frac{[(\beta_i-\beta_j)^2+(\delta_i-\delta_j)^2]x^2+2[\beta_j(\beta_i-\beta_j)+\delta_j(\delta_i-\delta_j)]x+(\beta_j^2+\delta_j^2)}{(|\lambda_i|^2-|\lambda_j|^2)x+|\lambda_j|^2}.$$

The minimum of $f(z)$ is obtained by setting $f'(x) = 0$. Omitting the details, we ascertain that the numerator of $f'(x)$ is

$$(\beta_i^2 + \delta_i^2 - \beta_j^2 - \delta_j^2)[(\beta_i - \beta_j)^2 + (\delta_i - \delta_j)^2]x^2$$
$$+ 2(\beta_j^2 + \delta_j^2)[(\beta_i - \beta_j)^2 + (\delta_i - \delta_j)^2]x$$
$$+ 2(\beta_j^2 + \delta_j^2)[\beta_j(\beta_i - \beta_j) + \delta_j(\delta_i - \delta_j)]$$
$$- (\beta_i^2 + \delta_i^2 - \beta_j^2 - \delta_j^2)(\beta_j^2 + \delta_j^2),$$

which further reduces to

(3.6-8) $\qquad N(x) = (|\lambda_i|^2 - |\lambda_j|^2)x^2 + 2|\lambda_j|^2 x - |\lambda_j|^2.$

The zeros of the latter are $x^{\pm} = |\lambda_j|/(|\lambda_j| \pm |\lambda_i|)$. If the root $|\lambda_j|/(|\lambda_j|-|\lambda_i|)$ is negative, it is of course not acceptable as $|z_i|^2$, and if it is positive, then $1 - x$ is negative and not acceptable as $|z_j|^2$. If it is zero, we are in the previous case of a single nonzero component.

Hence

(3.6-9) $\qquad |z_i|^2 = \dfrac{|\lambda_j|}{|\lambda_j| + |\lambda_i|} \quad \text{and} \quad |z_j|^2 = \dfrac{|\lambda_i|}{|\lambda_j| + |\lambda_i|},$

as claimed in the theorem. For this first total antieigenvector z, $f(z)$ may be written as

$$f(z) = \frac{((\beta_i - \beta_j)x + \beta_j)^2 + ((\delta_i - \delta_j)x + \delta_j)^2}{(|\lambda_i|^2 - |\lambda_j|^2)x + |\lambda_i|^2}.$$

Substituting the root $x = |\lambda_j|(|\lambda_j| + |\lambda_i|)^{-1}$ into this expression yields (after some simplification)

(3.6-10) $\qquad |\mu_1|^2(T) = f(z) = \dfrac{(\beta_i|\lambda_j| + \beta_j|\lambda_i|)^2 + (\delta_i|\lambda_j| + \delta - j|\lambda_i|)^2}{(|\lambda_i| + |\lambda_j|)^2 |\lambda_i||\lambda_j|}.$

Hence, the first total eigenvector $|\mu_1|(T)$ is the square root of the right-hand side of (3.6-10).

Now suppose that exactly three components, say, z_i, z_j, z_k, are nonzero at a minimizing point z^0 on the unit sphere. We may proceed to rule out this possibility. For such a z^0, $f(z)$ is of the form

$$f(z^0) = \frac{(\beta_i|\lambda_i^0|^2+\beta_j|\lambda_j^0|^2+\beta_k|z_k^0|^2)^2+(\delta_i|z_i^0|^2+\delta_j|z_j^0|^2+\delta_k|z_k^0|^2)^2}{|\lambda_i|^2|z_i^0|^2+|\lambda_j|^2|z_j^0|^2+|\lambda_k|^2|z_k^0|^2}.$$

For general three component, $z = (0, \ldots, z_i, 0, \ldots, 0, z_j, 0, \ldots, 0, \ldots)$, let $|z_i|^2 = v_1, |z_j|^2 = v_2, |z_k|^2 = v_3$, and consider the simplex,

(3.6-11) $\qquad V: v_1 + v_2 + v_3 = 1, \quad 0 \leq v_1 \leq 1, \; 0 \leq v_2 \leq 1, \; 0 \leq v_3 \leq 1.$

3.6 Antieigenvalue Theory 71

Let R be the minimum of the positive function $g(v_1, v_2, v_3) \equiv \sqrt{f(z)}$ on the simplex V. We may suppose that this minimum is attained at a strictly interior point (v_1^0, v_2^0, v_3^0) in V. But then it may be seen that we are assured of a minimizing point on the boundary of V as well. Moreover, by the convexity of the function $g(v_1, v_2, v_3)$, the minimum attained at the boundary point is strictly smaller than that at the interior, contradicting the presence of three nonzero components in the first place. Similarly, four and higher numbers of components may be ruled out.

Since R is the smallest positive number such that the graph of $g(v_1, v_2, v_3) - R$ touches one point of the simplex, then the intersection of the graph with one coordinate plane, say, the (v_i, v_j) plane, must cut the side of the simplex in that plane. This implies that

$$(3.6\text{-}12) \quad \inf_{\substack{v_i+v_j=1 \\ 0 \leq v_i \leq 1 \\ 0 \leq v_j \leq 1}} \frac{\sqrt{(\beta_i v_i + \beta_j v_j)^2 + (\delta_i v_i + \delta_j v_j)^2}}{\sqrt{|\lambda_i|^2 v_i + |\lambda_j|^2 v_j}} = R_1 \leq R,$$

and consequently we have

$$\frac{\sqrt{(\beta_i v_i + \beta_j v_j + \beta_k v_k)^2 (\delta_i v_i + \delta_j v_j + \delta_k v_k)^2}}{\sqrt{|\lambda_i|^2 v_i + |\lambda_j|^2 v_j + |\lambda_j|^2 v_k}} \geq R_1$$

for any (v_i, v_j, v_k) with $v_i + v_j + v_k = 1$, $0 \leq v_i \leq 1$, $0 \leq v_j \leq 1$, $0 \leq v_k \leq 1$. This convexity argument thus demonstrates that no more than two nonzero components need to be considered for the antieigenvector. □

Theorem 3.6-3. *For a normal operator as in Theorem 3.6-2, all higher total antieigenvalues take their values from the set $G \cup \{1\}$, and all corresponding higher total antieigenvectors possess the same two-component structure given in Theorem 3.6-2.*

Proof. The arguments all dealt with minimizing the positive function $f(z)$ on the unit sphere $\sum_{i=1}^{n} |z_i|^2 = 1$. The only property of the z_i used was the expression of $|(Tz, z)|/\|Tz\| \cdot \|z\|$ in terms of them. The convexity argument ruling out three or more components in a minimizing vector took place completely in terms of those three z_i and did not use any other normality properties of T. □

Another approach (Mirman, 1983, see Notes) proposed some interesting convexity methods for the calculation of the antieigenvalues of accretive operators. The numerical range $W(S)$ of an auxiliary operator S plays a fundamental role in these methods.

Theorem 3.6-4. *Let T be strictly accretive, $S = \operatorname{Re} T + iT^*T$, $\xi_1 + i\eta_1$, $\xi_2 + i\eta_2 \in \overline{W(S)}$, $\xi_1 \leq \xi_2$; define*

$$\tilde{\xi} = \frac{\xi_1 \eta_2 - \xi_2 \eta_1}{\eta_2 - \eta_1}.$$

72   3. Operator Trigonometry

Then $\mu_1(T) \le \tilde{\mu}$, where
(a) if $\eta_1 < \tilde{\eta}_2$, $\xi_1/2 < \tilde{\xi} < \xi_2/2$, then

$$\tilde{\mu}^2 = 4\tilde{\xi}\frac{\xi_2 - \xi_1}{\eta_2 - \eta_1};$$

(b) $\tilde{\mu}_2 = \xi_1^2/\eta_1$ if $\eta_1 \ge \eta_2$ or if $\eta_1 < \eta_2$ and $\tilde{\xi} \le \xi_1/2$;
(c) $\tilde{\mu}^2 = \xi_2^2/\eta_2$ if $\eta_1 < \eta_2$ and $\tilde{\xi} \ge \xi_2/2$.

Proof. One may verify that

(3.6-13) $\qquad \mu_1^2(T) = \inf\{\xi^2/\eta : \xi + i\eta \in W(S)\}.$

Hence, the problem reduces to finding the minimum of $J(\xi, \eta)$ on the line segment joining $\xi_1 + i\eta_1$ and $\xi_2 + i\eta_2$. If this minimum occurs at either end point of the line segment, then its value is either ξ_1^2/η_1 or ξ_2^2/η_2.

If it occurs at an interior point, then at that point the line segment and $J[\xi, \eta]$ should have parallel gradients. The equation of the line segment is

$$\eta - \eta_1 = \frac{\eta_2 - \eta_1}{\xi_2 - \xi_1}(\xi - \xi_1)$$

or

$$\eta - \eta_1 + \frac{\eta_1 - \eta_2}{\xi_2 - \xi_1}\xi + \xi_1\frac{\eta_2 - \eta_1}{\xi_2 - \xi_1} = 0.$$

Therefore, we should have for some ζ

$$\frac{2\xi}{\eta} = \zeta\frac{\eta_1 - \eta_2}{\xi_2 - \xi_1}$$

and

$$-\frac{\xi^2}{\eta^2} = \zeta,$$

which implies

$$\xi = 2\frac{\xi_1\eta_2 - \xi_2\eta_1}{\eta_2 - \eta_1} \quad \text{and} \quad \eta = \frac{\xi_1\eta_2 - \xi_2\eta_1}{\xi_2 - \xi_1},$$

and therefore

$$\frac{\xi^2}{\eta} = 4\frac{(\xi_1\eta_2 - \xi_2\eta_1)(\xi_2 - \xi_1)}{(\eta_2 - \eta_1)^2}.$$

Taking

$$\tilde{\xi} = \frac{\xi_1\eta_2 - \xi_2\eta_1}{\eta_2 - \eta_1},$$

we have

$$\frac{\xi^2}{\eta} = 4\tilde{\xi}\frac{\xi_2 - \xi_1}{\eta_2 - \eta_1}.$$

3.6 Antieigenvalue Theory

Hence, the minimum of ξ^2/η on the line segment is one of the values ξ_1^2/η_1, ξ_2^2/η_2, and $4\tilde{\xi}((\xi_2 - \xi_1)/(\eta_2 - \eta_1))$. One can actually find out which of the three values is the smallest using conditions (a), (b), or (c) in the theorem. □

Let us give one further result of this type. The proof combines convexity methods and the Lagrange multiplier approach.

Theorem 3.6-5. *Let T be strictly accretive. If $\|T\| < \infty$, then*

(3.6-14) $$\mu_1^2(T) = 4\max_t(\lambda_t t) > 0,$$

where λ_t is the lower bound of the spectrum of the operator

(3.6-15) $$S_t = \operatorname{Re} T - tT^*T.$$

Proof. Let $S = \operatorname{Re} T + iT^*T$, $\operatorname{Re} T$, and $\operatorname{Im} S = T^*T$. From (3.6-13) we have

$$\mu_1^2(T) = \inf\{\xi^2/\eta : \xi + i\eta \in W(S)\}.$$

Let ℓ_t be a straight line of support of $W(S)$, with $W(S)$ to the right of ℓ_t; then λ_t is the lower bound of $\sigma(\operatorname{Re} T - tT^*T)$. Therefore, we should consider the minimum of the function $J(\xi, \eta) = \xi^2/\eta$ on such lines. The equation of such a line is $\eta = (1/t)(\xi - \lambda_t)$ or $\eta - (1/t)\xi + (1/t)\lambda_t = 0$, and hence we must have

$$\frac{2\xi}{\eta} = \zeta\left(-\frac{1}{t}\right), \qquad \frac{-\xi^2}{\eta^2} = \zeta,$$

from which we have $\xi = 2\lambda_t$ and $\eta = \lambda_t/t$, and so $\xi^2/\eta = 4\lambda_t^2/(\lambda_t/t) = 4t\lambda_t$. □

At this point, it is useful to consider some simple examples of antieigenvalues and antieigenvectors. from which we have $\xi = 2\lambda_t$ and $\eta = \lambda_t/t$ and so

Example. Consider the 2×2 matrix

$$A = \begin{bmatrix} 9 & 0 \\ 0 & 16 \end{bmatrix}, \qquad \begin{aligned} \lambda_1 &= m = 9, \\ \lambda_2 &= M = 16. \end{aligned}$$

We know that all antieigenvectors are of the form

$$\left\{\frac{\pm\sqrt{\lambda_j}}{\sqrt{\lambda_i + \lambda_j}}, \frac{\sqrt{\lambda_i}}{\sqrt{\lambda_i + \lambda_j}}\right\},$$

where the indices i and j run through all eigenvalues. There are only a few combinations for this example, from which immediately one finds

$$z_1 = \text{1st antieigenvector} = (-4/5, 3/5).$$

74   3. Operator Trigonometry

The angle that it turns A is seen from
$$\frac{\langle Az_1, z_1 \rangle}{\|Az_1\|} = \frac{2(12^2)}{5^2 \cdot 12} = \frac{24}{25} \cos A = \mu_1(A) = 0.96.$$

The second antieigenvector, if required to be orthogonal to z_1, is then seen to be
$$z_2 = \text{2nd antieigenvector} = (3/4, 4/5),$$
which turns A by an angle whose cosine is
$$\frac{\langle Az_2, z_2 \rangle}{\|Az_2\|} = \frac{(3)(27) + (4)(64)}{\sqrt{(27)^2 + (64)^2}} = \frac{337}{347.3110997} = \mu_2(A) = 0.97031.$$

Thus, the angle $\phi(A)$ is $16.260°$ and the second critical angle $\phi_2(A)$ would be $13.997°$. These angles are easily visualized by plotting the first antieigenvector $z_1 = (-4, 3)$ and its image $Az_1 = (-3, 4)$; similarly, $z_2 = (3, 4)$ and its image $Az_2 = (27, 64)$, in 2-space.

Notice, however, that
$$z_1' = (4/5, 3/5)$$
is also a first antieigenvector. Generally, antieigenvectors come in pairs, are not orthogonal, and linear combinations of them are not antieigenvectors. This has led us to an alternate formulation of the higher antieigenvalues and corresponding antieigenvectors, namely, a combinatorial theory of higher antieigenvalues corresponding to the critical angles obtained from higher antieigenvectors constructed from systematically deleted sets of eigenvectors of the matrix. This will be illustrated by a numerical example in the next chapter.

Another Example. Consider the 3×3 matrix
$$A = \begin{bmatrix} 1 & 0 & 0 \\ 0 & 1 & 0 \\ 0 & 0 & 2 \end{bmatrix}, \quad \begin{matrix} \lambda_1 = 1, \\ \lambda_2 = 1, \\ \lambda_3 = 2. \end{matrix}$$

The formula for $\cos A$ for selfadjoint A immediately gives
$$\mu_1(A) = \cos(A) = \frac{2\sqrt{2}}{3} = 0.94281.$$

It is also easily seen as above that both
$$z_1 = (0, \sqrt{2}, 1),$$
$$z_2 = (0, -\sqrt{2}, 1)$$
are first antieigenvectors. Note that they are *not* orthogonal. If we still persist in asking that higher antieigenvectors be orthogonal to prior antieigenvectors, we are left in this example with the third antieigenvector
$$z_3 = (1, 0, 0),$$

which does not turn A at all and is in fact an eigenvector.

The full theory of higher antieigenvalues and their corresponding higher antieigenvectors, especially for arbitrary matrices and operators, is yet to be worked out. Perhaps several different versions will be needed, as dictated by the operator classes and applications considered.

Let us turn back to the first antieigenvalue μ_1 and derive a fundamental result for it and its corresponding antieigenvector pair. For simplicity, we consider only the case A bounded, strongly accretive. Consider the antieigenvalue functional

$$\mu(u) = \operatorname{Re} \frac{\langle Au, u \rangle}{\|Au\|\|u\|}.$$

Theorem 3.6-6 (Antieigenvector equation). *The Euler equation for the antieigenvalue functional $\mu(u)$ is*

(3.6-16)
$$2\|Au\|^2\|u\|^2(\operatorname{Re} A)u - \|u\|^2 \operatorname{Re}\langle Au, u\rangle A^*Au - \|Au\|^2 \operatorname{Re}\langle Au, u\rangle u = 0.$$

Scalar multiples of solutions are solutions, but the solution space is generally not a subspace. When A is normal, the Euler equation is satisfied not only by the first antieigenvectors but also by all of the eigenvectors of A.

Proof. To find the Euler equation, we consider the quantity

$$\left.\frac{d\mu}{dw}\right|_{\epsilon=0} = \lim_{\epsilon \to 0} \frac{\dfrac{\operatorname{Re}\langle A(u+\epsilon w), u+\epsilon w\rangle}{\langle A(u+\epsilon w), A(u+\epsilon w)\rangle^{1/2}\langle u+\epsilon w, u+\epsilon w\rangle^{1/2}} - \dfrac{\operatorname{Re}\langle Au, u\rangle}{\langle Au, Au\rangle^{1/2}\langle u, u\rangle^{1/2}}}{\epsilon}.$$

Let the expression on the right hand side be denoted $R_A(u, w, \epsilon)$. We have

$$\epsilon R_A(u, w, \epsilon) = [\operatorname{Re}\langle Au, u\rangle + 2\epsilon \operatorname{Re}\langle (\operatorname{Re} A)u, w\rangle$$
$$+ \epsilon^2 \langle Aw, w\rangle]\langle Au, Au\rangle^{1/2}\langle u, u\rangle^{1/2}/D$$
$$- [\langle Au, Au\rangle + 2\epsilon \operatorname{Re}\langle Au, Aw\rangle$$
$$+ \epsilon^2 \langle Aw, Aw\rangle]^{1/2}[\langle u, u\rangle + 2\epsilon \operatorname{Re}\langle u, w\rangle$$
$$+ \epsilon^2 \langle w, w\rangle]^{1/2} \operatorname{Re}\langle Au, u\rangle/D,$$

where D is the common denominator

$$D = \langle A(u+\epsilon w), A(u+\epsilon w)\rangle^{1/2}\langle u+\epsilon w, u+\epsilon w\rangle^{1/2}\langle Au, Au\rangle^{1/2}\langle u, u\rangle^{1/2}.$$

At this point, in deriving the Euler equations for eigenvalues of a selfadjoint operator, one gets a fortuitous cancellation of the ϵ-independent terms in the expression analogous to $\epsilon R_A(u, w, \epsilon)$, and the Euler equation immediately follows. Although that fortuitous situation does not occur here, we may attempt to mimic it by expanding the two square root bracket

76   3. Operator Trigonometry

expressions of the second numerator term

$$[\langle Au, Au\rangle + x(\epsilon)]^{1/2} = \langle Au, Au\rangle^{1/2} + \frac{1}{2}\langle Au, Au\rangle^{-1/2}x(\epsilon)$$
$$- \frac{1}{8}\langle Au, Au\rangle^{-3/2}x^2(\epsilon) + \cdots$$

and

$$[\langle u, u\rangle + y(\epsilon)]^{1/2} = \langle u, u\rangle^{1/2} + \frac{1}{2}\langle u, u\rangle^{-1/2}y(\epsilon)$$
$$- \frac{1}{8}\langle u, u\rangle^{-3/2}y^2(\epsilon) + \cdots,$$

where $x(\epsilon)$ and $y(\epsilon)$ are the ϵ-dependent terms, respectively, and where ϵ is sufficiently small relative to $\langle Au, Au\rangle$ and $\langle u, u\rangle$, respectively. Then we obtain $\langle Au, Au\rangle^{1/2}\langle u, u\rangle^{1/2}\mathrm{Re}\,\langle Au, u\rangle$ term cancellations, from which

$$\epsilon R_A(u, w, \epsilon) = \frac{[2\epsilon\mathrm{Re}\langle(\mathrm{Re}\,A)u, w\rangle + \epsilon^2\langle Aw, w\rangle]\langle Au, Au\rangle^{1/2}\langle u, u\rangle^{1/2}}{D}$$
$$- \frac{\mathrm{Re}\langle Au, u\rangle[\langle u, u\rangle^{1/2}r(\epsilon) + \langle Au, Au\rangle^{1/2}s(\epsilon)]}{D}.$$

where $r(\epsilon)$ and $s(\epsilon)$ denote the remainder terms in the square root series expansions above. To be specific,

$$r(\epsilon) = \frac{1}{2}\langle Au, Au\rangle^{-1/2}x(\epsilon) - \frac{1}{8}\langle Au, Au\rangle^{-3/2}x^2(\epsilon) + \cdots,$$
$$s(\epsilon) = \frac{1}{2}\langle u, u\rangle^{-1/2}y(\epsilon) - \frac{1}{8}\langle u, u\rangle^{-3/2}y^2(\epsilon) + \cdots,$$

where

$$x(\epsilon) = 2\epsilon\mathrm{Re}\,\langle Au, Aw\rangle + \epsilon^2\langle Aw, Aw\rangle,$$
$$y(\epsilon) = 2\epsilon\mathrm{Re}\,\langle u, w\rangle + \epsilon^2\langle w, w\rangle.$$

We may now divide by ϵ, from which

$$D \cdot R_A(u, w, \epsilon) = [2\mathrm{Re}\,\langle(\mathrm{Re}\,A)u, w\rangle + \epsilon\langle Aw, w\rangle]\|Au\|\|u\|$$
$$- \mathrm{Re}\,\langle Au, u\rangle[\|Au\|^{-1}\|u\|\mathrm{Re}\,\langle Au, Aw\rangle + O(\epsilon)]$$
$$- \mathrm{Re}\,\langle Au, u\rangle[\|u\|^{-1}\|Au\|\mathrm{Re}\,\langle u, w\rangle + O(\epsilon)].$$

Note also that by the above expansions,

$$D = [\|Au\| + O(\epsilon)][\|u\| + O(\epsilon)]\|Au\|\|u\| \to \|Au\|^2\|u\|^2$$

as $\epsilon \to 0$.

Thus, in the $\epsilon \to 0$ limit of $R_A(u, w, \epsilon)$, we arrive at

$$\left.\frac{d\mu}{dw}\right|_{\epsilon=0} =$$
$$\frac{2\mathrm{Re}\,\langle(\mathrm{Re}\,A)u, w\rangle\|Au\|^2\|u\|^2 - \mathrm{Re}\,\langle Au, u\rangle[\|u\|^2\mathrm{Re}\,\langle Au, Aw\rangle + \|Au\|^2\mathrm{Re}\,\langle u, w\rangle]}{\|Au\|^3\|u\|^3}.$$

3.6 Antieigenvalue Theory

Setting this expression to zero yields

$$2\|Au\|^2\|u\|^2\mathrm{Re}\langle(\mathrm{Re}A)u,w\rangle - \mathrm{Re}\langle Au,u\rangle[\|u\|^2\mathrm{Re}\langle A^*Au,w\rangle$$
$$+ \|Au\|^2\mathrm{Re}\langle u,w\rangle] = 0,$$

for arbitrary w, and hence the Euler equation

$$2\|Au\|^2\|u\|^2(\mathrm{Re}\,A)u - \|u\|^2\mathrm{Re}\langle Au\rangle A^*Au - \|Au\|^2\mathrm{Re}\langle Au,u\rangle u = 0. \quad \square$$

As the Euler equation is homogeneous (of order 5), scalar multiples of solutions are solutions. This was immediate from the variational quotient (3.6-1) for μ, but as it turns out the implications are greater from the Euler equation (3.6-16): for selfadjoint or normal operators, all eigenvectors also satisfy it. The details of the verification are left to the reader.

The following example is useful to see what is going on here.

Example. Let

$$A = \begin{bmatrix} 2 & 0 \\ 0 & 1 \end{bmatrix} \quad \text{and} \quad x = (x_1, x_2) = (x_1, \lambda x_1),$$

with λ real. Then $\|x\| = (1+\lambda^2)^{1/2}|x_1|$, $\langle Ax, x\rangle = (2+\lambda^2)|x_1|^2$, and $\|Ax\| = \langle 4+\lambda^2\rangle^{1/2}|x_1|$, so the antieigenvalue functional $\mu(x)$ is

$$F(\lambda) = \frac{\mathrm{Re}\,\langle Ax, x\rangle}{\|Ax\|\|x\|} = \frac{(2+\lambda^2)}{(\lambda^4+5\lambda^2+4)^{1/2}}.$$

Let us also consider the usual Rayleigh quotient

$$Q(\lambda) = \frac{\langle Ax, x\rangle}{\langle x, x\rangle} = \frac{2+\lambda^2}{1+\lambda^2}.$$

Then, for large $\lambda \to \infty$, we have $F(\lambda) \to 1$, $Q(\lambda) \to 1$, 1 being the smaller eigenvalue of A, and $x \to (0,1)$, the corresponding eigenvector. For small $\lambda \to 0$, we have $F(\lambda) \to 1$, $Q(\lambda) \to 2$, i.e., to the larger eigenvalue of A, and $x \to (1,0)$, the corresponding eigenvector. As we know from our general theory, $F(\lambda)$ attains its minimum $\mu_1 = \frac{2}{3}\sqrt{2} = 0.9428090416$ at the first antieigenvectors $x = (\pm 1, \sqrt{2})$, i.e., at $\lambda = \pm\sqrt{2}$. Checking this against $F(\lambda)$ above, we have $F(\pm\sqrt{2}) = 4/(4+10+4)^{1/2} = 0.9428090416$.

Calculating the derivative, we have

(3.6-17)
$$F'(\lambda) = \frac{(\lambda^4+5\lambda^2+4)^{1/2}(2\lambda)-(2+\lambda)^2(2^{-1})(\lambda^4+5\lambda^2+4)^{-1/2}(4\lambda^3+10\lambda)}{\lambda^4+5\lambda^2+4}$$
$$= \frac{(\lambda^4+5\lambda^2+4)(4\lambda)-(2+\lambda)^2(4\lambda^3+10\lambda)}{2(\lambda^4+5\lambda^2+4)^{3/2}}$$
$$= \frac{\lambda(\lambda^2-2)}{(\lambda^4+5\lambda^2+4)^{3/2}}.$$

Thus $F'(\lambda) = 0$ at $\lambda = 0$ and $\lambda = \pm\infty$, corresponding to eigenvectors, and at $\lambda = \pm\sqrt{2}$, corresponding to antieigenvectors.

Notes and References for Section 3.6

The terminology "antieigenvalues" and "antieigenvectors" was introduced by

K. Gustafson (1972). "Antieigenvalue Inequalities in Operator Theory," in *Inequalities* III, Proceedings Los Angeles Symposium, 1969, ed. O. Shisha, Academic Press, 115–119.

Upper bounds for $\mu_1(T)$ for T a finite-dimensional, strongly accretive normal operator were obtained in

C. Davis (1980), "Extending the Kantorovich Inequalities to Normal Matrices," *Linear Algebra Appl.* **31**, 173–177.

The antieigenvalues and antieigenvectors for finite-dimensional normal operators, Theorems 3.6-1 to 3.6-3, were obtained by

K. Gustafson and M. Seddighin (1989), "Antieigenvalue Bounds," *J. Math. Anal. Appl.* **143**, 327–340.

K. Gustafson and M. Seddighin (1993), "A Note on Total Antieigenvectors," *J. Math. Anal. Appl.* **178**, 603–611.

The convexity methods using the numerical range $W(S)$, where $S = \operatorname{Re} T + iT^*T$, giving Theorem 3.6-4, were developed by

B. Mirman (1983), "Antieigenvalues: Method of Estimation and Calculation," *Linear Algebra Appl.* **49**, 247–255.

The Euler equation of Theorem 3.6-6 was known and casually mentioned at the 1969 Inequality III Symposium (Gustafson, 1972), but the proof was not published in full until

K. Gustafson (1995), "Antieigenvalues," *Linear Algebra Appl.* **208/209**, 437–454.

Because both antieigenvectors and eigenvectors, in the selfadjoint and normal operator cases, satisfy this Euler equation, this theory constitutes a significant extension of the Rayleigh–Ritz variational theory of eigenvectors.

In the paper just mentioned, the new, combinatorial selection theory of eigenvectors, especially higher antieigenvectors, was formulated. This is advanced further by a computational example in

K. Gustafson (1995), "Matrix Trigonometry," *Linear Algebra Appl.* **217**, 117–140.

That example shows a combinatorial selection of higher antieigenvalues (i.e., smaller critical angles) developing during an iterative calculation.

Also in the latter paper, the minmax equality theorem (3.2-1) is revisited, with the proof now indicating a two-component nature for antieigenvectors for general (strongly accretive) operators. How much of that two-component structure can be represented in terms of eigenvectors remains to be further explored.

Some further exposition of the antieigenvalue theory may be found in

K. Gustafson (1996). *Lectures on Computational Fluid Dynamics, Mathematical Physics, and Linear Algebra*, Kaigai Publications, Tokyo.

Endnotes for Chapter 3

Other good names for *antieigenvalues* and *antieigenvectors* would be *anglevalues* and *anglevectors*, *turning angles* and *turning vectors*, and so on. Perhaps, in deference to the German *eigenvalue* and *eigenvector*, the eventual names should be *winkelvalue* and *winkelvector*. Winkel means "angle" or "corner" in German, angle becomes *vinkel* in Swedish and Danish, but in Dutch winkel becomes *shop*, presumably descending from "the shop on the corner"! Indeed, more descriptive names for eigenvalue and eigenvector would be *dilationvalue* and *dilationvector*, or their equivalent in another language such as German or French (e.g., in the former, *streckenvalue* or *dehnenvalue*)? Surely, we should accept evolutionary precedents and the terms "eigenvalue" and "eigenvector", and that is why the same precedent evolved the terms "antieigenvalue" and "antieigenvector". Moreover, as we have seen in this chapter, the latter in some ways are composed of the former, again reinforcing the importance of accepting the priority of the term *eigen* ($\approx$ self, inherent) to the values and vectors to which it has become historically attached.

4
Numerical Analysis

Introduction

The fact that numerical range and numerical analysis carry the same adjective is a historical accident. Indeed, numerical range derives from the German *Wertvorrat*, meaning value field, and as we saw in Chapter 1, it originated as the continuous range of bilinear forms. Numerical analysis, on the other hand, usually connotes a conversion from continuous to discrete and then to a computation yielding precise numbers.

Nonetheless, there are interesting connections between numerical range and numerical analysis—most importantly, application of the former to the latter. The spectrum of an operator underlies many parts of numerical analysis, and that spectrum is contained within the numerical range of that operator. To the extent of this common tie of both the numerical range and numerical analysis to the spectrum, we can expect fruitful interplay.

It turns out, for example, that the convergence rate of the standard numerical analysis steepest descent algorithm is exactly $\sin A$. This rather beautiful new result connecting numerical analysis and numerical range carries over to the conjugate gradient algorithm, where the convergence rate is seen to be $\sin(A^{1/2})$. After presenting these recent results, we go back to the general numerical analysis of initial-value problems and examine the important Von Neumann–Kreiss stability theorem in terms of the numerical range. Then we give a very short treatment of the role of numerical range considerations in computational fluid dynamics, both from the finite-difference and finite-element points of view. From there we turn to the workhorse Lax–Wendroff scheme for hyperbolic conservation laws, seeing that its stability condition is that of a numerical range contained in the unit disk. Then the recent extension of that theory to spectral methods and the corresponding notion of pseudo-eigenvalues is presented.

4.1 Optimization Algorithms

The method of *steepest descent* is one of the basic numerical methods in optimization theory. In steepest descent for the minimization of a function f, a basic algorithm is

(4.1-1) $$x_{k+1} = x_k - \alpha_k \nabla f(x_k)^T.$$

If we restrict attention to the quadratic case, where

$$f(x) = \frac{\langle x, Ax \rangle}{2} - \langle x, b \rangle,$$

where A is a symmetric, positive definite matrix with eigenvalues $0 < m = \lambda_1 \leq \lambda_2 \leq \cdots \leq \lambda_n = M$, then the point of minimum x^* solves the linear system

$$Ax^* = b.$$

For the quadratic minimization (i.e., the linear solver problem $Ax = b$), the descent algorithm becomes

(4.1-2) $$x_{k+1} = x_k - \frac{\|y_k\|^2 y_k}{\langle Ay_k, y_k \rangle},$$

where $y_k = Ax_k - b$ is called the residual error. Letting

$$E_A(x) = \frac{\langle (x - x^*), A(x - x^*) \rangle}{2} = f(x) + \frac{\langle x^*, Ax^* \rangle}{2}$$

measure the error in the iterates, one has the fundamental Kantorovich error bound

$$E_A(x_{k+1}) \leq \left(1 - \frac{4\lambda_1 \lambda_n}{(\lambda_n + \lambda_1)^2}\right) E_A(x_k).$$

But in terms of $\lambda_1 = m$ and $\lambda_n = M$ and the operator trigonometry of Chapter 3, this becomes

$$\begin{aligned} E_A(x_{k+1}) &\leq (1 - \mu_1^2(A)) E_A(x_k) \\ &= (1 - \cos^2 A) E_A(x_k) \\ &= (\sin^2 A) E_A(x_k). \end{aligned}$$

Thus the error rate of the method, in the $A^{1/2}$ norm $E_A(x)^{1/2}$, is exactly $\sin A$.

Theorem 4.1-1 (Trigonometric convergence). *In quadratic steepest descent, for any initial point x_0, there holds at every step k*

(4.1-3) $$E_A(x_{k+1}) \leq (\sin^2 A) E_A(x_k).$$

Proof. This follows from the discussion above and Theorem 3.1-2. □

We may interpret this result geometrically as follows. The first antieigenvalue $\mu_1(A) \equiv \cos A$ measures the maximum turning capability of A. Thus the angle $\phi(A)$ is a fundamental constraint on iterative methods involving A. Steepest descent does a good job but cannot converge faster than the maximum distance from x to Ax, which is represented trigonometrically after normalization by the quantity $\sin A$.

More generally, in optimization theory for nonquadratic problems, one uses the Hessian of the objective function to guide descent by quadratic approximation. The smallest and largest eigenvalues m and M of the Hessians at each iteration point then determine the convergence rate of the method. Thus under rather general conditions, (e.g., objective function $f(x, \ldots, x_n)$ with continuous second partial derivatives and a local minimum to which you are converging), the objective values $f(x_k)$ converge to the minimum linearly with convergence rate bounded by $\sin H$ of the Hessian.

In (Luenberger (1984); see the Notes) it is shown how one may use such convergence theory to compare methods. Let us observe that the use of $\sin A$ can sharpen such comparisons.

Example. One approach to minimizing f is to solve the equations of the necessary local minimization condition $\nabla f(x) = 0$. It has been proposed to apply steepest descent to the function $h(x) = |\nabla f(x)|^2$. Considering for simplicity only the case of quadratic $f(x) = \frac{1}{2}\langle x, Ax \rangle - \langle b, x \rangle$, for which $\nabla f = Ax - b$, it may be seen that this means that steepest descent is to be applied to the function

$$h(x) = \langle x, A^2 x \rangle - 2\langle x, Ax \rangle + \|b\|^2.$$

But we know by Theorem 4.1-1 that steepest descent applied to this quadratic function h will have convergence rate governed by $\sin(A^2)$. This will always be a slower convergence than that which would accrue simply by using steepest descent in the first place.

The convergence ratio of the two methods may be compared for large *condition number* κ, which in this case is $\kappa = M/m$. By the approximations

(4.1-4)
$$\left(\frac{\kappa-1}{\kappa+1}\right)^2 \cong (1 - \kappa^{-1})^4,$$
$$\left(\frac{\kappa^2-1}{\kappa^2+1}\right)^2 \cong (1 - \kappa^{-2})^4,$$
$$\left(1 - \kappa^{-2}\right)^\kappa \cong 1 - \kappa^{-1},$$

it follows that it takes about κ steps of the proposed method to match one step of ordinary steepest descent. In terms of our operator trigonometric theory, the comparison is more explicit. The ratio of convergence rates in the A^2 inner product to that in the A inner product is

(4.1-5) $$\frac{\sin(A^2)}{\sin A} = \frac{(M+m)^2}{M^2+m^2} = 1 + \frac{2mM}{M^2+m^2} = 1 + \cos(A^2).$$

Example. Let A be any symmetric square matrix with $m = 1/4$ and $M = 1$. Then A^2 has lower bound $m^2 = 1/16$ and upper bound $M^2 = 1$. For these two matrices A and A^2, we thus have

$$\cos A = 4/5 = 0.8,$$
$$\sin A = 3/5 = 0.6,$$
$$\cos A^2 = 8/17 = 0.47059,$$
$$\sin A^2 = 15/17 = 0.88235.$$

Thus the angles of these operators are

$$\phi(A) \cong 36.8699°,$$
$$\phi(A^2) \cong 61.9275°.$$

The convergence rates are thus

$$E_A(x_{k+1}) \leqq 0.36\, E_A(x_k)$$

and

$$E_{A^2}(x_{k+1}) \leqq 0.77554\, E_A(x_k),$$

respectively.

Notes and References for Section 4.1

The sin A steepest descent trigonometric convergence rate result Theorem 4.1-1 was first presented in
K. Gustafson (1991). "Antieigenvalues in Analysis," *Proceedings of the Fourth International Workshop in Analysis and its Applications*, June 1–10, 1990, Dubrovnik, Yugoslavia, eds. C. Stanojevic and O. Hadzic, Novi Sad, Yugoslavia, 57–69.
Further details are given in
K. Gustafson (1994). "Operator Trigonometry," *Linear and Multilinear Algebra* **37**, 139–159.
For optimization methods and steepest descent methods, see
D. Luenberger (1984). *Linear and Nonlinear Programming*, 2nd Edition, Addison–Wesley.
   The angle of an operator $\phi(A)$ of Section 3.1 is *not* the same notion as the angle between subspaces, which one finds in the invariant subspace theory, e.g., see
G. Stewart and J. G. Sun (1990). *Matrix Perturbation Theory*, Academic Press, Boston,

84   4. Numerical Analysis

for an account of this theory, which goes back to Krein, Krasnoselski, Milman, Paige, Kato, Davis, Kahan, Wielandt, and others. The operator angle $\phi(A)$ is really a two-dimensional concept, focusing attention on the *most noninvariant* one-dimensional subspaces!

In the same vein, the fundamental new geometrical understanding of the Kantorovich bound given in the trigonometric convergence theorem (Theorem 4.1-1) is *not* the same as the Kantorovich–Wielandt inequality: an excellent summary of the Kantorovich–Wielandt inequality is given in [HJ1], which we discovered only during the writing of this book. The angle θ in the Kantorovich–Wielandt inequality is motivated entirely by the condition number $\kappa = \lambda_n/\lambda_1$ and is defined by $\cot(\theta/2) = \lambda_n/\lambda_1$. Then, the Kantorovich–Wielandt inequality states that

$$\langle Ax, Ay \rangle \leqq \cos\theta \|Ax\| \|Ay\|$$

for all pairs of mutually orthogonal vectors x and y. The geometrical interpretation of this inequality is that the smallest angle between Ax and Ay is at most θ and hence θ means the minimum attainable angle between Ax and Ay as x and y range over all orthonormal pairs of vectors.

However, during this writing we have found that a wonderful, heretofore unnoticed connection between this angle θ and our $\phi(A)$ indeed exists: $\cos\phi(A^2) = \sin\theta$! This connection between two quite distinct geometrical perspectives is elaborated in
K. Gustafson (1996c). "The Geometrical Meaning of the Kantorovich–Wielandt Inequalities," to appear.

4.2 Conjugate Gradient

Steepest descent algorithms are sometimes slow to converge. By contrast, conjugate gradient methods have the advantage of converging, if one ignores roundoff, in N iterations, for an $N \times N$ symmetric positive definite matrix A. Similar to the theory for steepest descent, one knows that in the A inner product error measure

$$E_A(x) = \frac{\langle (x - x^*), A(x - x^*) \rangle}{2}.$$

the conjugate gradient error rate is governed by (see Luenberger, 1984; Section 4.1 References)

$$E_A(x_k) \leq 4 \left(\frac{\sqrt{\kappa(A)} - 1}{\sqrt{\kappa(A)} + 1} \right)^{2k} E_A(x_0)$$

for any initial guess x_0. Remembering that the condition number $\kappa(A) = M/m$, we may rewrite this as

$$\|x_k - x^*\|_{A^{1/2}} \leqq 2 \left(\frac{M^{1/2} - M^{1/2}}{M^{1/2} + m^{1/2}} \right)^k \|x_0 - x^*\|_{A^{1/2}},$$

from which the following result ensues.

Theorem 4.2-1. *For A a positive-definite symmetric matrix, for any initial guess x_0, the conjugate gradient iterates x_k converge to the solution x^* of $Ax = b$ with error rate*

(4.2-1) $\qquad \|x_k - x^*\|_{A^{1/2}} \leqq 2(\sin(A^{1/2}))^k \|x_0 - x^*\|_{A^{1/2}}$

Proof. This follows from the above discussion and the fact that the spectrum $\sigma(A^{1/2}) = (\sigma(A))^{1/2}$ by the spectral mapping theorem. Recall that for selfadjoint T one knows that $\sin T = (M_T - m_T)/(M_T + m_T)$ □

Example. We reverse the example of the previous section. Let A be any real symmetric, positive definite $N \times N$ matrix with largest eigenvalue $\lambda_{\max} = M = 1$ and smallest eigenvalue $\lambda_{\min} = m = 1/16$. Then the steepest descent algorithm has convergence rate governed by the $\sin A$ error bound of Theorem 4.1-1, namely

$$E_A(x_{k+1}) \leqq 0.77854 \, E_A(x_k),$$

whereas the increased efficiency of the conjugate gradient algorithm is governed by the $\sin A^{1/2}$ bound of Theorem 4.2-1, namely

$$E_A(x_{k+1}) \leqq 0.36 \, E_A(x_k).$$

Let us now look at some numerical computations on simple examples.

Example. Let

$$A = \begin{bmatrix} 1 & 0 \\ 0 & 2 \end{bmatrix},$$

for which we know $\cos A = 0.9428090416$ and the angle $\phi(A) = 19.47122063°$
Let us solve $Ax = b$,

$$b = \begin{pmatrix} 1 \\ 1 \end{pmatrix},$$

first by steepest descent and then by conjugate gradient. The solution

$$x^* = \begin{pmatrix} 1 \\ 0.5 \end{pmatrix}$$

will be sought from initial guess

$$x_0 = \begin{pmatrix} 0 \\ 0 \end{pmatrix}.$$

Here are the results for steepest descent, to five iterations:

x		$E_A(x_n)$	$\mu_1(x_n)$	$\phi(x_n)$
0	0			
0.6667	0.6667	0.08333	0.94868	18.435
0.8889	0.4444	0.00926	0.94868	18.435
0.963	0.5185	0.00103	0.94653	18.820
0.9877	0.4938	0.00011	0.94868	18.435
0.9959	0.5021	0.00001	0.94842	18.482

Here $\mu_1(x_n) = \langle Ax_n, x_n \rangle / \|Ax_n\| \|x_n\|$ measures the cosine of the angle of each iteration. This leads us to remark that the error $E_A(x_n)$ decreases in very near correspondence to its sin A trigonometric convergence rate estimate. The angle of each iteration is close to $\phi(A)$. Such behavior has been observed previously in steepest descent, but here it is seen trignometrically, i.e., in terms of the numerical range trignometric theory.

Using the conjugate gradient algorithm algorithm, we find

x		$E_A(x_n)$	$\mu_1(x_n)$	$\phi(x_n)$
0	0			
0.2941	0.5882	0.25692	0.97619	12.529
1	0.5	3.5e -15	0.94868	18.435

Next, we consider a less trivial example.

Example. Let

$$A = \begin{bmatrix} 20 & 0 & 0 & 0 \\ 0 & 10 & 0 & 0 \\ 0 & 0 & 2 & 0 \\ 0 & 0 & 0 & 1 \end{bmatrix}.$$

From $\sin A = 19/21$, we have the angle $\phi(a) = 64.7912347°$ Let us solve $Ax = b = (1,1,1,1)$, for which the solution $x^* = (0.05, 0.1, 0.5, 1)$ is sought from initial guess $x_0 = (0,0,0,0)$.

Steepest descent converges very slowly. To achieve an error of 10^{-6}, 64 iterations were required. We leave most of them out, showing only the beginning and final values in the following table.

x				$E_A(x_n)$	$\phi(x_n)$
0	0	0	0		
0.1212	0.1212	0.1212	0.1212	0.5826	42.757
0.0078	0.1043	0.1815	0.1912	0.4464	48.550
0.1030	0.0994	0.2533	0.2824	0.3464	57.141
0.0180	0.0999	0.2929	0.3399	0.2710	47.236
0.0905	0.1000	0.3399	0.4148	0.2133	57.029
...	...	...	...	...	...
0.0501	0.1000	0.5000	0.9985	$1.40e-06$	42.758
0.4994	0.1000	0.5000	0.9987	$1.14e-06$	42.793
0.4994	0.1000	0.5000	0.9987	$9.26e-07$	42.758

Again, we see error decreasing like the $\sin^2 A = 0.8186$ trigonometric convergence estimate. However, the angle of each iteration is less than $\phi(A)$, as it must also accommodate critical subangle effects due to higher antieigenvectors other than those composed of just the first and last eigenvectors.

Using the conjugate gradient algorithm, we find

x				$E_A(x_n)$	$\phi(x_n)$
0	0	0	0		
0.0594	0.0297	0.0059	0.0030	0.7667	13.526
0.0498	0.1063	0.0265	0.0133	0.7112	21.526
0.0500	0.09997	0.5901	0.3062	0.2488	41.676
0.04995	0.1000	0.5000	1.0000	$2.9e-08$	42.745
0.0500	0.1000	0.5000	1.0000	$2.9e-10$	42.758

Again, we note the departure from the extreme angle $\phi(A)$ to what we may call a mixed convergence angle.

Notes and References for Section 4.2

The $\sin(A^{1/2})$ conjugate gradient convergence rate was first given by
K. Gustafson (1994). "Operator Trigonometry," *Linear and Multilinear Algebra* **37**, 139–159.
The examples above were taken from
K. Gustafson (1994). "Antieigenvalues," *Linear Algebra Appl.* **208/209**, 437–454.
K. Gustafson (1995). "Matrix Trigonometry," *Linear Algebra Appl.* **217**, 117–140.
The 4 × 4 example is the one we referred to in the previous Section 3.6, showing the emergence of combinatorially selected higher antieigenvectors.

New trigonometric interpretations of many computationally important recent variants of the conjugate gradient algorithm, such as GCR, GCR(k), GCG, PCG, Orthomin, CGN, GMRES, CGS, BCG, and others, will be found in
K. Gustafson (1996). "Trigonometric Interpretation of Iterative Methods," in *Proc. Conf. on Algebraic Multilevel Iteration Methods with Applications*, (O. Axelsson, ed.), June 13-15, Nijmegen, Netherlands, 23–29.

It should be remarked that the angle $\phi(A)$ is continuous (for invertible A, and in the operator norm $\|A\|$) in A and therefore so are its trigonometric functions. Thus, trigonometric criteria for numerical schemes can generally be expected to be stable under small perturbations (e.g., computational roundoff errors in the elements of A or in iteration matrices related to A). In this sense, the fundamental role of the operator angle $\phi(A)$ in iterative algorithms reproduces the same virtue as that of the numerical range $W(A)$, perturbation stability.

4.3 Discrete Stability

In Section 3.4, we briefly mentioned the abstract theory of initial-value problems

(4.3-1)
$$\begin{cases} \dfrac{du(t)}{dt} = Au & t > 0, \\ u(0) = u_0. \end{cases}$$

If A is an m-dissipative operator, then the solution $u(t)$ is generally given by the semigroup $u(t) = e^{At}u_0$ for u_0 in the domain of A. Important instances of this theory occur in ordinary differential equations, partial differential equations, quantum mechanics, and various engineering disciplines.

When one wishes a numerical solution of such problems, one must approximate the continuous problem (4.3-1) by a discrete one. By this we mean that instead of evolving forward over continuous time $t > 0$, the solution will evolve forward in discrete time steps Δt. Also, the operator A must be approximated in a finite way to enable concrete computation. This is accomplished in the finite-difference method by replacing A by a discrete version defined on a specified finite grid of points x_i.

A classic example is the heat equation

(4.3-2)
$$\begin{cases} \dfrac{\partial u}{\partial t} = \dfrac{\partial^2 u}{\partial x^2}, \\ u(0) = f(x) \text{ given.} \end{cases}$$

A standard easy discretization is the forward Euler (in time), central difference (in space) approximation:

(4.3-3)
$$\dfrac{\partial u}{\partial t} \simeq \dfrac{u(x, t + \Delta t) - u(x, t)}{\Delta t},$$
$$\dfrac{\partial^2 u}{\partial x^2} \simeq \dfrac{u(x + \Delta x, t) - 2u(x, t) + u(x - \Delta x, t)}{(\Delta x)^2}.$$

Given appropriate boundary conditions to uniquely determine the solution, the solution is then calculated by propagating forward over the discrete $(\Delta x, \Delta t)$ lattice in space-time. Labeling the x points on this lattice $x_i = i\Delta x$ and the t points $t_j = j\Delta t$, and calling the discrete system solution $U_{i,j} = U(x_i, t_j)$, the propagation algorithm becomes

(4.3-4) $$U_{i,j+1} = U_{i,j} + \dfrac{\Delta t}{(\Delta x)^2} [U_{i_1,j} - 2U_{i,j} + U_{i-1,j}].$$

This scheme is usually called the *Euler explicit scheme*.

There are better schemes than the Euler scheme, but it is conceptually simple and illustrates the basic ideas, so it will serve us here. We will ignore boundary-value considerations and just look at the numerical solutions to such a discretized initial-value problem. These solutions evolve forward in time as a sequence of matrix iterations.

For example, we may write the explicit Euler scheme (4.3-4) more simply as

(4.3-5) $$U^{j+1} = C(\Delta t)U^j,$$

where

(4.3-6) $$C(\Delta t) = I + (\Delta t)A_D$$

and where A_D denotes the discrete centered difference approximation to $\partial^2 u/\partial x^2$ in (4.3-3). $C(\Delta t)$ plays the role of the semigroup for the discrete problem, just as $U(\Delta t) = e^{A\Delta t}$ was the semigroup for the continuous problem. Note that $C(\Delta t)$ is the linear (first two terms) approximation to the formal Taylor series expansion of $e^{A_D \Delta t}$. If we look at

$$C(\Delta t)C(\Delta t)u = (I + 2\Delta t A_D + (\Delta t)^2 A_D^2)u$$
$$= C(2\Delta t)u + (\Delta t)^2 A_D^2 u,$$

we can see indeed that to first order $C(\Delta t)$ is a semigroup.

Recalling from (3.4-2) that one recovers the infinitesimal generator from the semigroup by the strong limit of the difference quotient as $t \to 0$, which for the heat equation would be

(4.3-7) $$Au \equiv \frac{\partial^2 u}{\partial x^2} = s - \lim_{\Delta t \to 0} \frac{(e^{A\Delta t} - I)}{\Delta t} u$$

for all u in $D(A)$, and noting that from (4.3-6) we may write

(4.3-8) $$A_D u = \frac{C(\Delta(t)) - I}{\Delta t} u,$$

we may expect that if we have done a good job of discretizing the original problem, then

(4.3-9) $$\|(A_D - A)u(t)\| \to 0 \quad \text{as} \quad \Delta t \to 0$$

for all solutions $u(t)$ of (4.3-1) for all good initial data $u_0(x)$. This is called the *consistency condition* for a discretization: that on all true solutions the discretized operator approximates the continuous one with truncation error going to zero as the discretization tends to the continuum. Most natural discretizations have this property, although often it is not easy to prove it.

Given the consistency or assumed consistency of the discrete equation, one then comes to one of the most important concepts in the numerical analysis of linear initial-value problems, the *Lax equivalence theorem*: the discrete solutions converge to the continuous solution iff the scheme is stable. In one sense, this is a numerical version of the *Banach theorem*: T is onto iff $(T^*)^{-1}$ is bounded. However, in numerical as well as the analytical settings, you must define your "stability" carefully, and then prove that the Banach or Lax "theorems" actually hold for the operators and stability of interest.

Stability comes in many versions, but in our present setting it means that the successive iterates of the operators $C(\Delta T)$ of the scheme are uniformly bounded,

(4.3-10) $$\|(C(\Delta t))^n\| \leq M$$

for all small $0 < \Delta t \leq \tau$, on the interval of interest $0 \leq n\Delta t \leq T$, $n = 1, 2, 3, \ldots$. Otherwise, we might fear that the discrete solution would blow up in finite time as the grid size goes to zero. On the other hand, merely restricting the iteration matrices to be bounded in (4.3-10) may at first seem too weak to guarantee convergence. However, the consistency condition (4.3-9) is strong, and with additional assumptions of smoothness of solutions, one can usually use specific properties of the discretized scheme to bound what are called local discretization errors and then, using those combined with (4.3-10), prove the convergence of the method. Conversely, by use of the *uniform boundedness principle*, one can show that convergence of all discrete solutions U_D to u implies (4.3-10).

We may now reconnect our discussion to the numerical range. There are essentially two (related) connections within the present context. The first is the power-boundedness theorem (Theorem 2.1-1), whereby we know that if the numerical radius $w(C) \leq 1$, then the same is true for all powers, i.e., $w(C^n) \leq 1$. Since the numerical radius is equivalent to the operator norm, a stability such as that desired in (4.3-10) can be demonstrated. This connection and potential application of numerical range to numerical stability led to the considerable theoretical research interest in power-bounded operator families in the 1960s, and thus to the mapping theorems of Chapter 2.

The second connection to the numerical range in such applications comes from going to the energy norm. For our purposes, we shall regard this as going from the operator norm $\|A\|$ to the numerical range norm $w(A)$. In applications it often comes down to an appropriate integration by parts. It would take us too far afield here to fully elaborate this connection, but let us outline it for one important instance, that of the *Von Neumann–Kreiss stability theorem*.

By substituting Fourier series for U^{j+1} and U^j in the iteration (4.3-5), one converts the problem to

(4.3-11) $$V^{j+1}(\mathbf{k}) = G(\Delta t, \mathbf{k}) V^j(\mathbf{k}),$$

where $V(\mathbf{k})$ denotes the Fourier coefficients, $\mathbf{k}$ being the vector of frequencies within the Fourier expansion. The matrices $G(\Delta t, \mathbf{k})$ are called the (Von Neumann) amplification matrices, and for stability of the original scheme, a uniform bound

(4.3-12) $$\|(G(\Delta t, \mathbf{k}))^n\| \leq K$$

is desired, for all $0 < \Delta t < \tau$, $0 \leq n\Delta t \leq T$, and all $\mathbf{k}$. The stability theorem is the following.

Theorem 4.3-1 (Discrete boundedness). *Equivalently:*
(1) *A family of square matrices* $\{A\}$ *is uniformly power bounded*
$$\|A^n\| \leq M$$
for some M, *all* $n = 1, 2, 3, \ldots$, *and all* A *in* $\{A\}$.
(2) *The family is uniformly resolvent bounded*
$$\|(zI - A)^{-1}\| \leq K(|z| - 1)^{-1}$$
for some K, *all* $|z| > 1$, *and all* A *in* $\{A\}$.
(3) *The vector space* V *on which the family* $\{A\}$ *operates can be uniformly transformed to* W, $w = T_A v$, *on which the transformed* A *has norm*
$$\|T_A A T_A^{-1}\| \leq 1,$$
where the transforms T_A *are uniform in the sense that* $T_A^* T_A = H_A$, *a Hermitian positive definite matrix satisfying*
$$A^* H_A A \leqq H_A$$
and
$$C^{-1} \leqq H_A \leqq C$$
for some C, *all* H_A, *and all* A *in* $\{A\}$.

Proof. See (Richtmyer and Morton (1967); see Notes). The equivalence of the uniform power-boundedness (1) and the uniform resolvent-boundedness (2) can be reasonably expected from faith in the functional calculus for such operator families. The renorming (3) can be seen in terms of the numerical range if we imagine only the case $T = T^* = H^{1/2}$, for then from $A^* H A \leqq H$ we have
$$H^{-1/2} A^* H^{1/2} H^{1/2} A H^{-1/2} \leq 1$$
or $\|H^{1/2} A H^{-1/2}\|^2 \leqq 1$. This is nothing more than the fact that the numerical radius and the operator norm are the same for selfadjoint operators T^*T. □

Theorem 4.3-1 is important because it can then be shown that for a number of finite-difference schemes, stability becomes
$$\|U^{j+1}\|_H \leq (1 + O(\Delta t))\|U^j\|_H,$$
which is satisfied for (4.3-6) if

(4.3-13) $$\|C(\Delta t)\|_H \leqq 1 + O(\Delta t).$$

This puts the burden of proof for stability on showing that A_D is $O(1)$ in (4.3-6). For $A_D u$ the finite difference of the one-dimensional Laplacian $-\Delta_1 u = -\partial^2 u/\partial x^2$ in the heat equation example (4.3-2), that implies that we should consider the energy norm $\langle A_D u, u \rangle = \langle \operatorname{grad} u, \operatorname{grad} u \rangle$ for (the square of) our H-norm, for any operator is $O(1)$ in its own operator norm.

In this way, the discrete boundedness theorem (Theorem 4.3-1) can be seen as yet another instance of the benefits in going from strong to weak formulation in the theory of partial differential equations.

Notes and References for Section 4.3

For discrete initial-value problems, see
R. D. Richtmyer and K. Morton (1967). *Difference Methods for Initial Value Problems*, Wiley, New York.
As pointed out there, when amplification matrices are normal operators, then the spectral, numerical and operator radii are all equal, and we can conclude stability by knowing that all eigenvalues are contained within the closed unit disk. Generally, the amplification matrices should not be expected to be normal operators when treating complicated systems of partial differential equations and specified boundary conditions, although it sometimes does happen.

The discrete boundedness theorem (Theorem 4.3-1) was first shown by H. O. Kreiss (1962). "Über die Stabilitätsdefinition für Differenzengleichungen die partielle Differentialgleichungen approximieren," *BIT* **2**, 153–181. When A is indeed normal and with spectrum $\sigma(A)$ within the unit disk, then $A - zI$ and $(A - zI)^{-1}$ are also normal, and by Theorem 1.4-3, for $|z| > 1$,

$$\|(A - zI)^{-1}\|^{-1} = d(z, \sigma(A)) \geq |z| - 1$$

so that condition (2) of Theorem 4.3-1 holds with constant $K = 1$. Such considerations can be taken as motivation for the larger class of normal-like operators and growth conditions, in which K may be larger, to be treated in Chapter 6.

A return to the power-boundedness of the $N \times N$ matrices in Theorem 4.3-1 and in particular to the question of the best M and K in (1) and (2), respectively, and their relationship, may be found in
R. J. Le Veque and L. N. Trefethen (1984). "On the Resolvent Condition in the Kreiss Matrix Theorem," *BIT* **24**, 585–591.
M. N. Spijker (1991). "On a Conjecture by Le Veque and Trefethen Related to the Kreiss Matrix Theorem," *BIT* **31**, 551–555.

For purposes of tighter estimates on the stability constants of finite-difference schemes, another variant on the numerical range $W(A)$, called the M-numerical range, was introduced in
H. W. J. Lenferink and M. N. Spijker (1990). "A Generalization of the Numerical Range of a Matrix," *Linear Algebra Appl.* **140**, 251–266.
This numerical range is based on growth rates of the operator powers $\|(A - \gamma I)^k\|$; see Section 5.5 for further details. The M-numerical range is a generalization of the algebraic numerical range of $[BD]$ and nicely brings

4.4 Fluid Dynamics

Fluid dynamics is a vast field that has greatly influenced the development of mathematics and, in particular, numerical analysis, operator theory, and matrix methods. The *Navier–Stokes* equations for the velocities u,

(4.4-1)
$$\begin{cases} \dfrac{du}{dt} = \underbrace{k(u)\Delta u}_{\text{diffusion}} - \underbrace{u \cdot \nabla u}_{\text{convection}} - \underbrace{\nabla p}_{\substack{\text{pressure} \\ \text{gradients}}}, & t > 0, \\ u(0) = u_0, \end{cases}$$

may be regarded as an initial-value problem, as in Section 3.4, although now with a nonlinear term added, and its discretizations lead to considerations similar to those of Section 4.3. However, its multidimensionality and vector nature, along with its nonlinearity and boundary conditions, have led to a whole new field of study, commonly now called computational fluid dynamics.

For our purposes here of illustrating how numerical range considerations, chiefly positive definiteness, enter into the numerical analysis of fluid dynamics, let us consider the one-dimensional model problem

(4.4-2)
$$\begin{cases} u_t + v u_x - k u_{xx} = 0, & 0 < x < 1, \ t > 0, \\ u(x,0) = f(x), \text{ given} & 0 < x < 1, \\ u(0,t) = u(1,t) = 0, & t > 0. \end{cases}$$

An initial velocity distribution $u(x,0)$ is to evolve forward in time, subject to zero boundary conditions at the ends of the fluid domain $0 \leq x \leq 1$. The reader may prefer to imagine u to be a temperature, as in the heat equation of the previous section, rather than a velocity. However, we now have a convective term vu_x balanced against a diffusion term ku_{xx}. We have linearized the problem by taking a constant given positive convection speed v and a constant positive viscosity k.

The finite-difference method proceeds as in the previous section. The Euler explicit discretizations (4.3-3) now yield the discrete equation

(4.4-3)
$$\frac{U_{i,j+1} - U_{i,j}}{\Delta t} + v \frac{U_{i,j} - U_{i,j-1}}{\Delta x} - k \left[\frac{U_{i-1,j} - 2U_{i,j} + U_{i+1,j}}{(\Delta x)^2} \right] = 0.$$

We may write this, analogous to (4.3-4), as

(4.4-4)
$$U_{i,j+1} = \left[\frac{k+v\Delta x}{(\Delta x)^2}\right] U_{i-1,j} + \left[1 - \frac{2k+v\Delta x}{(\Delta x)^2}\Delta t\right] U_{i,j}$$
$$+ \left[\frac{k\Delta t}{(\Delta x)^2}\right] U_{i+1,j}.$$

It is easily checked that $U_{i,j+1}$ is a convex combination of the three preceding values, $U_{i-1,j}, U_{i,j}, U_{i+1,j}$, provided that

(4.4-5)
$$\frac{\Delta t}{(\Delta x)^2} \leq \frac{1}{2k+v\Delta x}.$$

This is a typical limitation on time step Δt one encounters when using Euler explicit discretizations, which may be overcome by going to so-called implicit schemes, which we will not discuss here. Accepting this time step condition, it follows that the $U_{i,j}$ discrete solutions are uniformly bounded independent of Δx and Δt. This is the case because, as a convex combination of the three nearest discrete values at the preceding time step, $U_{i,j+1}$ is bounded by the largest and smallest of those, which are bounded by the largest and smallest of all discrete values at the preceding time step. Thus, all are bounded eventually by the extremes of the initial data. This uniform boundedness stability, in the sense of (4.3-10), can then, with some more work, be used to prove rigorously the convergence of the discrete solutions to the true solution as the grids become arbitrarily fine.

The numerical range enters into many numerical analysis considerations, often in a rather elementary way: the positive definiteness of the discretized infinitesimal generators A_D. This property was called m-accretive in Section 3.4. The numerical range positive definiteness entered into Theorem 4.3-1 through the uniform condition $C^{-1} \leq H_A$ there. Without further belaboring this point, let us just assert that many numerical iterative methods need a numerical range with positive real part in order to assure convergence of the algorithm.

When encountering a given discretization, it can be instructive to write it on the coarsest grid possible in order to see its basic properties.

Example. Consider the problem (4.4-2) under scheme (4.4-3) on the grid $h = \Delta x = 1/3$. Disregarding the time derivative for the moment, let U_1 denote $u(1/3)$ and U_2 denote $u(2/3)$. Then, the steady problem (time derivatives set to zero) yields the 2×2 system

$$\frac{v(U_1 - U_2)}{h} - \frac{k(U_0 - 2U_1 + U_2)}{h^2} = 0, \quad j = 1,$$
$$\frac{v(U_2 - U_1)}{h} - \frac{k(U_1 - 2U_2 + U_3)}{h^2} = 0, \quad j = 2,$$

Letting $P = hv/k$, multiply by h^2/k, from which

(4.4-6) $\underbrace{\begin{bmatrix} 2 & -1 \\ -1 & 2 \end{bmatrix} \begin{bmatrix} U_1 \\ U_2 \end{bmatrix}}_{A_d \;=\; \text{diffusion}} + \underbrace{P \begin{bmatrix} 1 & 0 \\ -1 & 1 \end{bmatrix} \begin{bmatrix} U_1 \\ U_2 \end{bmatrix}}_{A_c \;=\; \text{convection}} = \underbrace{\begin{bmatrix} (1+P)U_0 \\ U_3 \end{bmatrix}}_{\text{Boundary data}}.$

Equation (4.4-6) shows the discrete balance of diffusion and convection that makes up the discrete infinitesimal generator $A_D = A_d + PA_c$. Both A_d and A_c are (real) positive definite:

(4.4-7) $\langle x, A_d x \rangle = 2(x_1^2 - x_1 x_2 + x_2^2) \geq \|x\|^2,$
$\langle x, A_c x \rangle = x_1^2 - x_1 x_2 + x_2^2 \geq \frac{1}{2}\|x\|^2,$

and hence A_D is positive definite with lower bound $m(A_D) = 1 + P/2$. This foretells that the discrete time-dependent problem

(4.4-8) $\dfrac{du}{dt} = \left(-\dfrac{k}{h^2} A_d + \dfrac{v}{h} A_c\right) u$

corresponding to (4.4-2) will exponentiate well, i.e., that the problem is well posed with unique solution. The number P is a discrete Reynolds number for the problem, often called the Peclet number. If we let the basic given convective velocity v become negative, when the Peclet number $P = -2$, the lower bound $m(A_D)$ passes through zero and we can expect numerous numerical difficulties. For the coarse $h = 1/3$ grid of this example, that is when the convective velocity is -6 times the basic diffusion constant k. In this instance, a numerical remedy is simply to "downwind," i.e., difference the convection term in the opposite direction. This creates what is then called an implicit scheme.

Next, let us briefly consider the finite-element method. We will see similar numerical range considerations, notably positive definiteness, enter in a critical way. Assuming zero boundary conditions, the partial differential equation of (4.4-2) may be multiplied by an arbitrary "test function" ϕ and integrated over the interval, from which, by an integration by parts,

(4.4-9) $\displaystyle\int_0^1 u_t \phi + v \int_0^1 u_x \phi + k \int_0^1 u_x \phi_x = 0.$

Let us assume a solution of the separated form

(4.4-10) $u(x,t) = \displaystyle\sum_{i=0}^{N} c_i(t)\phi_i(x),$

where the ϕ_i are some desirable linearly independent basis set. Substituting (4.4-10) into (4.4-9) and taking ϕ to be $\phi_1, \ldots, \phi_N$ in turn yields the so-called Galerkin $N \times N$ ordinary differential equation system of the form

(4.4-11) $M_D C'(t) + A_D C(t) = 0.$

96   4. Numerical Analysis

Assuming M_D to be invertible, this is now the discrete initial-value problem

(4.4-12)
$$\begin{cases} \dfrac{dC(t)}{dt} = -M_D^{-1} A_D C(t), \\ C(0) = c_0, \end{cases}$$

where c_0 is the vector of initial coefficients from the initial-value representation $u(x, 0) = \sum_{i=1}^{N} c_i(0) \phi_i(x)$.

The following coarse discretization illustrates this method.

Example. Let $N = 2$, and let $\phi_1(x)$ be the piecewise linear "hat" function which rises with slope $1/h$ from $x = 0$ to $x = h$, then descends with slope $-1/h$ from $x = h$ to $x = 2h$, and is 0 for x between $2h$ and $3h = 1$. Similarly, define $\phi_2(x)$ to be 0 from $x = 0$ to $x = h$, rising to 1 at $2h$, then down to 0 at $x = 3h = 1$.

In this discretization, the Galerkin finite-element system (4.4-11) becomes

(4.4-13)
$$\begin{bmatrix} \int_0^1 \phi_1 \phi_1 & \int_0^1 \phi_2 \phi_1 \\ \int_0^1 \phi_1 \phi_2 & \int_0^1 \phi_2 \phi_2 \end{bmatrix} \begin{bmatrix} c_1' \\ c_2' \end{bmatrix} + \begin{bmatrix} v \begin{bmatrix} \int_0^1 \phi_1' \phi_1 & \int_0^1 \phi_2' \phi_1 \\ \int_0^1 \phi_1' \phi_2 & \int_0^1 \phi_2' \phi_2 \end{bmatrix} \\ + k \begin{bmatrix} \int_0^1 \phi_1' \phi_1' & \int_0^1 \phi_2' \phi_1' \\ \int_0^1 \phi_1' \phi_2' & \int_0^1 \phi_2' \phi_2' \end{bmatrix} \end{bmatrix} \begin{bmatrix} c_1 \\ c_2 \end{bmatrix} = \begin{bmatrix} 0 \\ 0 \end{bmatrix}.$$

Doing the integrations yields

(4.4-14)

$$h \underbrace{\begin{bmatrix} 2/3 & 1/6 \\ 1/6 & 2/3 \end{bmatrix}}_{\text{mass matrix}} \begin{bmatrix} c_1' \\ c_2' \end{bmatrix} + v \underbrace{\begin{bmatrix} 0 & 1/2 \\ -1/2 & 0 \end{bmatrix}}_{\text{convection}} + \frac{k}{h} \underbrace{\begin{bmatrix} 2 & -1 \\ -1 & 2 \end{bmatrix}}_{\text{diffusion}} \begin{bmatrix} c_1 \\ c_2 \end{bmatrix} = \begin{bmatrix} 0 \\ 0 \end{bmatrix}$$

Note that the mass matrix M_D is positive definite and that we arrive at the same positive definite diffusion matrix A_d as before. The convection matrix A_c is no longer positive definite but has zero (real) numerical range. We remark that, by Theorem 2.4-1, the spectrum $\sigma(M_0^{-1} A_D)$ is real and positive. Thus, we expect that (4.4-14) will exponentiate well.

Finally, let us look at the equations of gas dynamics. There, in order to simplify, one assumes no viscosity k in the Navier–Stokes equations (4.4-1), but on the other hand one must allow a variable density. From these assumptions, one arrives at the important *Euler system* of three first-order partial differential equations:

(4.4-15)
$$\begin{pmatrix} \rho \\ \rho u \\ e \end{pmatrix}_t + \begin{pmatrix} \rho u \\ \rho u^2 + p \\ (e+p)u \end{pmatrix}_x = \begin{pmatrix} 0 \\ 0 \\ 0 \end{pmatrix}.$$

These are an example of a hyperbolic system of conservation laws. Here ρ is the density of the gas, $m = \rho u$ is the mass, u is the velocity, p is the pressure, $e = \rho\epsilon + \rho u^2/2$ is the total energy per unit volume, ϵ is the internal energy per unit mass, and only one space dimension x is considered.

Using an assumed additional equation of state $p = (\gamma-1)\rho\epsilon$, where $\gamma > 1$ is a known gas constant, and calculating the Jacobian of the system (4.4-15) to convert it to a quasilinear form, one arrives at the matrix system

(4.4-16)
$$\begin{pmatrix} \rho \\ m \\ e \end{pmatrix}_t + \begin{bmatrix} 0 & 1 & 0 \\ (\gamma-3)u^2/2 & (3-\gamma)u & \gamma-1 \\ (\Gamma-1)u^3 - \gamma eu/\rho & \gamma e/\rho - 3(\gamma-1)u^2/2 & \Gamma u \end{bmatrix} \begin{pmatrix} \rho \\ m \\ e \end{pmatrix}_x = \begin{pmatrix} 0 \\ 0 \\ 0 \end{pmatrix}.$$

The matrix A of (4.4-16) is an example of the nonnormal matrices one may encounter in the numerical analysis of gas dynamics. A simpler matrix may be obtained by writing (4.4-16) in terms of the primitive variables ρ, u, and p, from which one arrives at the system

(4.4-17)
$$\begin{pmatrix} \rho \\ u \\ p \end{pmatrix}_t + \begin{bmatrix} u & \rho & 0 \\ 0 & u & 1/\rho \\ 0 & \gamma p & u \end{bmatrix} \begin{pmatrix} \rho \\ u \\ p \end{pmatrix} = \begin{pmatrix} 0 \\ 0 \\ 0 \end{pmatrix}.$$

Of course, the matrix A of (4.4-17) depends on the unknown flow quantities themselves and hence will vary with them, but a qualitative theory may be obtained for such systems.

For example, one finds that the eigenvalues of A are $u, u+c$, and $u-c$, where $c^2 = \gamma p/\rho$ represents an acoustic propagation speed. From the corresponding eigenvectors, one may diagonalize A. But the theory is essentially incomplete for higher space dimensions. Therefore, the brunt of investigation has been computational. One such early scheme, the Lax–Wendroff scheme, will be examined in the next section.

It is interesting to ask what kind of operator the matrix A of (4.4-17) is, for the flow variables assumed constant at some designated time. A straightforward computation yields

(4.4-18)
$$AA^* - A^*A = \rho^2 \begin{bmatrix} 1 & 0 & c^2 \\ 0 & \rho^{-4}-1-c^4 & 0 \\ c^2 & 0 & c^4-\rho^{-4} \end{bmatrix}.$$

For fun, let us examine its eigenvalues. From the scalar equation $\det[AA^* - A^*A - \lambda I] = 0$, we arrive at

(4.4-19) $\qquad [\lambda^2 + (\rho^{-4} - c^4 - 1)\lambda - \rho^{-4}] \cdot [(\rho^{-4} - c^4 - 1) - \lambda] = 0.$

98   4. Numerical Analysis

Let b denote the term $(\rho^{-4} - c^4 - 1)$. Then, the eigenvalues are seen to be

(4.4-20) $$\lambda = b \quad \text{or} \quad \frac{-b \pm (b^2 + 4\rho^{-4})^{\frac{1}{2}}}{2}.$$

This shows the critical role of the evolving density ρ relative to the acoustic speed $c = (\gamma \rho / \rho)^{1/2}$ in determining the signs of the eigenvalues. For example, when $b \neq 0$, neither A or A^* is even hyponormal (see Chapter 6 for these operator classes). When $b = 0$, one has the interesting reduction that the eigenvalues $\lambda = 0, \pm \rho^{-2}$, depend only on the density.

Notes and References for Section 4.4

There are a great many references to fluid dynamics and its numerical analysis, but Richtmyer and Morton (1967) as cited in the preceding section remains an excellent starting treatment.

One may view discretization of partial differential equations as three approaches, the finite-difference, finite-element, and finite-spectral Methods, see
K. Gustafson (1987). *Partial Differential Equations*, 2nd Edition, Wiley, New York.
In the third editions of that book
K. Gustafson (1991, 1992). *Applied Partial Differential Equations I, II*, Kaigai Publishers, Tokyo, Japan (in Japanese).
K. Gustafson (1993). *Introduction to Partial Differential Equations and Hilbert Space Methods*, 3rd Edition, International Journal Services, Calcutta, India.
one will find treatment of hyperbolic systems of partial differential equations and computational gas dynamics.

4.5 Lax–Wendroff Scheme

If one takes a conservation law

(4.5-1) $$u_t + (F(u))_x = 0$$

such as (4.4-15) and differentiates it with respect to t, one gets an expression for u_{tt}, namely

(4.5-2) $$u_{tt} = -F_{tx} = -F_{xt} = -(Au_t)_x = (AF_x)_x,$$

where $A = A(u)$ is the Jacobian matrix of partial derivatives of F with respect to u. We may then substitute (4.5-1) and (4.5-2) into the Taylor series

(4.5-3)
$$U_{i,j+1} = U_{i,j} + \Delta t (U_t)_{i,j} + \frac{(\Delta t)^2}{2}(U_{tt})_{i,j} + \cdots$$
$$\cong U_{i,j} - \Delta t (F(U)_x)_{i,j} + \frac{(\Delta t)^2}{2}(AF(u_x)_x)_{i,j}$$

where we have gone from small u to big U to signal a discretization in progress. Assuming that A is constant, and using centered differences for the x derivatives, this gives the Lax–Wendroff second-order scheme

(4.5-4)
$$U_{i,j+1} = U_{i,j} - \frac{\Delta t}{2\Delta x} A(U_{i+1,j} - U_{i-1,j})$$
$$+ \frac{(\Delta t)^2}{2(\Delta x)^2} A^2 (U_{i+1,j} - 2U_{i,j} + U_{i-1,j}).$$

This is the basic *Lax–Wendroff* scheme, which provided a significant advance in the numerical treatment of gas dynamics. For the system (4.4-17), for example, the scheme (4.5-4) is stable provided that $\Delta t \leq \Delta x (|u| + c)^{-1}$, a fact rather easily obtained by the method of amplification matrix discussed in Section 4.3. Although the Lax–Wendroff scheme has been somewhat superseded by a number of newer schemes, it remains a basic workhorse scheme and its results a point of comparison for other schemes.

In this section, we will demonstrate the stability of a Lax–Wendroff scheme as an instance of the power inequality for the numerical range, Theorem 2.1-1.

Consider the first-order linear hyperbolic system in two space dimensions:

(4.5-5)
$$\begin{cases} u_t = A u_x + B u_y, & -\infty < x, y < \infty, \\ u(x,y,0) = f(x,y), \end{cases}$$

where $u = (u_1(x,y,t), u_2(x,y,t))$ is the unknown vector, A and B are given selfadjoint matrices, and u_x, u_y, u_t denote the usual partial derivatives. Let $\Delta x > 0$, $\Delta y > 0$ and $\Delta t > 0$ be the increments in a finite-difference scheme with fixed ratios

$$\lambda = \frac{\Delta t}{\Delta x} \quad \text{and} \quad \mu = \frac{\Delta t}{\Delta y}.$$

Taking the grid points to be

$$(x_j, y_k, t_m) = (j\Delta x, k\Delta y, m\Delta t), \quad m = 0, 1, 2, \ldots, \quad j, k = 0, \pm 1, \pm 2, \ldots,$$

and denoting $u(x_j, y_k, t_m)$ by u_{jk}^m, as described earlier, we may consider the Taylor expansion about (x_j, y_k, t_m)

$$u_{jk}^{m+1} = u_{jk}^m + \Delta t (u_t)_{jk}^m + \frac{1}{2}(\Delta t)^2 (u_{tt})_{jk}^m,$$

from which, as earlier, we arrive at the Lax–Wendroff discretization

(4.5-6)
$$\begin{aligned}v_{jk}^{m+1} &= v_{jk}^m + \frac{1}{2}\lambda A(v_{j+1,k}^m - v_{j-1,k}^m) + \frac{1}{2}\mu B(v_{j,k+1}^m - v_{j,k-1}^m) \\ &+ \frac{1}{2}\lambda^2 A^2(v_{j+1,k}^m - 2v_{j,k}^m + v_{j-1,k}^m) + \frac{1}{2}\mu^2 B^2(v_{j,k+1}^m \\ &\quad - 2v_{j,k}^m + v_{j,k-1}^m) \\ &+ \frac{1}{8}\lambda\mu(AB+BA)(v_{j+1,k+1}^m - v_{j+1,k-1}^m - v_{j-1,k+1}^m \\ &\quad + v_{j-1,k-1}^m).\end{aligned}$$

As discussed in Section 4.3, given that the scheme (4.5-6) is consistent, the question of its convergence reduces to that of its stability. To answer this question, we introduce the amplification matrix, obtained as in Section 4.3 by Fourier transform of the difference scheme:

(4.5-7)
$$\begin{aligned}T = T(\xi, \eta, \lambda, \mu) &= I + \frac{1}{2}\lambda A(e^{i\xi} - e^{-i\xi}) + \frac{1}{2}\mu B(e^{i\eta} - e^{-i\eta}) \\ &+ \frac{1}{2}\lambda^2 A^2(e^{i\xi} - 2 + e^{-i\xi}) + \frac{1}{2}\mu^2 B^2(e^{i\eta} - 2 + e^{-i\eta}) \\ &+ \frac{1}{8}\lambda\mu(AB+BA)(e^{i(\xi+\eta)} - e^{i(\xi-\eta)} - e^{i(\eta-\xi)} + e^{-i(\xi+\eta)}).\end{aligned}$$

We recall that the scheme (4.5-6) will be stable if there exists a norm $\|\ \|$ and a fixed constant $K > 0$ such that

$$\|(T^m)\| \leq K, \quad m = 1, 2, 3, \ldots, \quad -\pi \leq \xi \leq \pi, \ \pi \leq \eta \leq \pi.$$

Since all norms on M_n are equivalent, we can use the numerical radius $w(T)$.

Theorem 4.5-1. *The Lax–Wendroff scheme (4.5-6) is stable if*

(4.5-8a)
$$\lambda^2 w^2(A) + \mu^2 w^2(B) \leq \frac{1}{4},$$

or, equivalently, if the time step

(4.5-8b)
$$\Delta t \leq \frac{1}{2}\left[\frac{w^2(A)}{(\Delta x)^2} + \frac{w^2(B)}{(\Delta y)^2}\right]^{-1/2}.$$

Proof. Using the simplifications

$$e^{i\xi} - e^{-i\xi} = 2i\sin\xi, \qquad e^{i\xi} - 2 + e^{-i\xi} = 2\cos\xi - 2,$$

and

$$\begin{aligned}e^{i(\xi+\eta)} - e^{i(\xi-\eta)} - e^{i(\xi+\eta)} + e^{-i(\xi+\eta)} &= 2\cos(\xi+\eta) - 2\cos(\xi-\eta) \\ &= 4\sin\xi\sin\eta\end{aligned}$$

we can write T as $R + iJ$, where the selfadjoint operators R and J are given by

$$R = I - C$$

where

$$C = (1 - \cos\xi)\lambda^2 A^2 + (1 - \cos\eta)\mu^2 B^2 + \frac{1}{2}\lambda\mu \sin\xi \sin\eta(AB + BA)$$

and

$$J = \lambda \sin\xi A + \mu \sin\eta B.$$

We now prove that $w(T) \leq 1$. If so, then $w(T^n) \leq 1$ for all n by the numerical range power inequality of Chapter 2, and stability of the scheme (4.5-6) will have been shown.

To that end, we compute $\langle Tx, x\rangle$. For $x \in C^n$, $\|x\| = 1$,

(4.5-9)
$$\begin{aligned}|\langle Tx,x\rangle|^2 &= |\langle Rx,x\rangle|^2 + |\langle Jx,x\rangle|^2 \\ &= |\langle (I-C)x,x\rangle|^2 + |\langle Jx,x\rangle|^2 \\ &= 1 + \langle Cx,x\rangle^2 - 2\langle Cx,x\rangle + \langle Jx,x\rangle^2.\end{aligned}$$

We will prove that

(4.5-10) $\langle Jx,x\rangle^2 - 2\langle Cx,x\rangle \leq -(1-\cos\xi)^2\lambda^2\|Ax\|^2 - (1-\cos\eta)^2\|Bx\|^2\mu^2,$

which with (4.5-9) implies that
(4.5-11)
$$|\langle Tx,x\rangle|^2 \leq 1 + \langle Cx,x\rangle^2 - (1-\cos\xi)^2\lambda^2\|Ax\|^2 - (1-\cos\eta)\mu^2\|Bx\|^2.$$

Then, we will prove that

(4.5-12) $\qquad \langle Cx,x\rangle^2 \leqq (1-\cos\xi)^2\lambda^2\|Ax\|^2 + (1-\cos\eta)^2\mu^2\|Bx\|^2,$

which with (4.5-11) implies that

(4.5-13) $\qquad\qquad\qquad |\langle Tx,x\rangle|^2 \leqq 1.$

Turning first to (4.5-10), we see that

$$\begin{aligned}2C - J^2 &= 2(1-\cos\xi)\lambda^2 A^2 + 2(1-\cos\eta)\mu^2 B^2 \\ &\quad + \sin\xi \sin\eta \lambda\mu(AB+BA) \\ &\quad - [\lambda^2 \sin^2\xi A^2 + \mu^2 \sin^2\eta B^2 + \lambda\mu \sin\xi \sin\eta(AB+BA)] \\ &= \lambda^2(1-\cos\xi)^2 A^2 - \mu^2(1-\cos\eta)^2 B^2,\end{aligned}$$

and hence by the Schwarz inequality,

$$\begin{aligned}\langle Jx,x\rangle^2 - 2\langle Cx,x\rangle &\leq \langle (J^2 - 2C)x, x\rangle \\ &= -(1-\cos\xi)^2\lambda^2\|Ax\|^2 \\ &\quad - (1-\cos\eta)^2\mu^2\|Bx\|^2.\end{aligned}$$

Having established (4.5-10), we turn next to (4.5-12). To that end, we compute

(4.5-14)
$$\langle Cx, x\rangle = (1 - \cos\xi)\lambda^2\|Ax\|^2 + (1 - \cos\eta)\mu^2\|Bx\|^2$$
$$+ \frac{1}{2}\lambda\mu\sin\xi\sin\mu\langle(AB + BA)x, x\rangle.$$

Bounding the last term as

(4.5-15)
$$\frac{1}{2}|\lambda\mu\sin\xi\sin\eta\langle(AB + BA)x, x\rangle|$$
$$\leq |\lambda\mu\sin\xi\sin\eta\langle Ax, Bx\rangle|$$
$$\leq \lambda\mu|\sin\xi\sin\eta|\|Ax\|\|Bx\|$$
$$\leq \frac{1}{2}[\lambda^2\sin^2\xi\|Ax\|^2 + \mu^2\sin^2\eta\|Bx\|^2]$$
$$= \frac{1}{2}[\lambda^2(1 - \cos^2\xi)\|Ax\|^2 + \mu^2(1 - \cos^2\eta)\|Bx\|^2]$$
$$\leq (1 - \cos\xi)\lambda^2\|Ax\|^2 + (1 - \cos\eta)\mu^2\|Bx\|^2,$$

since we always have, for any θ,

$$\frac{1}{2}(1 - \cos^2\xi) = \frac{1}{2}(1 + \cos\theta)(1 - \cos\theta) \leq 1 - \cos\theta,$$

from (4.5-14) and (4.5-15), we thus have

$$|\langle Cx, x\rangle| \leq 2(1 - \cos\xi)\lambda^2\|Ax\|^2 + 2(1 - \cos\eta)\mu^2\|Bx\|^2.$$

By the Schwarz inequality,

$$\langle Cx, x\rangle^2 \leq 4[(1 - \cos\xi)^2\lambda^2\|Ax\|^2 + (1 - \cos\eta)^2\mu^2\|Bx\|^2][\lambda^2\|Ax\|^2$$
$$+ \mu^2\|Bx\|^2].$$

Since A and B are selfadjoint, we have

$$\|Ax\| \leq \|A\| = w(A) \quad \text{and} \quad \|Bx\| \leq w(B).$$

Therefore, using the condition $\lambda^2 w^2(A) + \mu^2 w^2(B) \leq \frac{1}{4}$ of Theorem 4.5-1, we have the desired (4.5-12). As noted earlier, this implies $w(T) \leq 1$ and $w(T^n) \leq 1$ by the power inequality. □

Notes and References for Section 4.5

The Lax–Wendroff scheme was first introduced in
P. Lax and B. Wendroff (1960). "Systems of Conservation Laws," *Comm. Pure Appl. Math.* **13**, 217–237.
The numerical range appears implicitly in their stability condition

$$|\langle u, Gu\rangle| \leq (1 + O(\Delta t))\|u\|^2$$

in
P. Lax and B. Wendroff (1964). "Difference Schemes for Hyperbolic Equations with High Order of Accuracy," *Comm. Pure Appl. Math.* **17**, 381–398.
Note that this is condition (4.3-13) although stated for the amplification matrix G rather than the scheme matrix C. Theorem 4.5-1 was shown in M. Goldberg and E. Tadmor (1982). "On the Numerical Radius and its Applications," *Linear Alg. Appl.* **42**, 263–284.

4.6 Pseudo Eigenvalues

The notion of *pseudo eigenvalues* has been recently introduced into numerical analysis to overcome a rather fundamental limitation encountered in actual computation: the sensitivity of eigenvalues to small perturbations. Many results in numerical theory are stated in terms of and depend rather precisely on the eigenvalues of operators. For selfadjoint and normal operators, or near normal operators, the eigenvalues are often insensitive to small perturbations. However, as one treats far from normal operators, extreme sensitivity can occur.

Example. Consider the left shift on n-space

$$(4.6\text{-}1) \qquad A = \begin{bmatrix} 0 & 1 & & & & 0 \\ & 0 & 1 & & & \\ & & 0 & 1 & & \\ & & & \ddots & & \\ & & & & \ddots & 1 \\ 0 & & & & & 0 \end{bmatrix}.$$

If one perturbs by B the $n \times n$ matrix of all zeros except its extreme lower left corner element b_{n1}, which is taken to be ϵ with ϵ small, then the eigenvalues of A (an n-fold 0) become the nth roots of $-\epsilon$. Thus, the single point spectrum $\sigma(A) = 0$ under perturbation has changed to n points on the circle of radius $\epsilon^{1/N}$. For example, for $n = 64$ and $\epsilon = 0.1$, the spectrum $\sigma(A + B)$ lies on the circle of radius 0.9646616199.

The fact that the numerical range is stable under additive perturbations, whereas the spectrum is not, i.e.,

$$\sigma(A + B) \not\subset \sigma(A) + \sigma(B),$$
$$W(A + B) \subset W(A) + W(B),$$

is one of the virtues of using the numerical range in perturbation contexts, especially for general operators or matrices for which the convex hull of the spectrum is significantly different from the numerical range. Between the spectrum $\sigma(A)$ and the *augmented numerical range*

$$(4.6\text{-}2) \qquad W_\epsilon(A) = W(A) + \Delta_\epsilon,$$

where Δ_ϵ denotes the closed disk of radius ϵ, is the *pseudo-spectrum* $\sigma_\epsilon(A)$, which is defined to be all λ that are the eigenvalues of some perturbed matrix $A + E$ with $\|E\| \leq \epsilon$.

Theorem 4.6-1. *Equivalently, for a given $n \times n$ matrix A, for given $\epsilon \geq 0$,*
  (i) λ *is an ϵ-pseudo-eigenvalue of A,*
  (ii) $\|(\lambda I - A)x\| \leq \epsilon$ *for some* $\|x\| = 1$,
  (iii) $\|(\lambda I - A)^{-1}\|^{-1} \leq \epsilon$,
  (iv) *the smallest singular value of $\lambda I - A$ is $\leq \epsilon$.*

Proof. We prove (i) $\Rightarrow$ (ii) $\Rightarrow$ (iii) $\Rightarrow$ (iv). If λ is an ϵ-pseudo-eigenvalue of A, then there is some perturbation E for which $(A+E)x = \lambda x$, $\|x\| = 1$. Thus $\|(\lambda I - A)x\| = \|Ex\| \leq \epsilon$. As $\|(\lambda I - A)^{-1}\|^{-1}$ is the infimum of $\|(\lambda I - A)x\|$ over all $\|x\| = 1$, (iii) follows immediately from (ii): here we have assumed that λ is not in $\sigma(A)$. Similarly, $\|(\lambda I - A)^{-1}\|^{-1}$ is the smallest singular value σ_n of $\lambda I - A$, by the spectral mapping theorem. Finally, given the singular value decomposition

$$(4.6\text{-}3) \qquad \lambda I - A = U\Sigma V^* = \sum_{j=1}^{n} \sigma_j u_j v_j^*.$$

we may take $E = \sigma_n u_n v_n^*$ so that

$$\lambda I - A - E = \sum_{j=1}^{n-1} \sigma_j u_j v_j^*,$$

which, being singular, means that λ is an eigenvalue of $A + E$. $\square$

The insensitivity of pseudo-spectra to perturbations, and the inclusion of the pseudo-spectra within the augmented numerical range, are perhaps the two basic properties of the pseudo-spectrum $\sigma_\epsilon(A)$, aside from the easily checked fact that it is closed.

Theorem 4.6-2 (Pseudo-spectra stability). *Let D be a perturbation of norm δ. Then*

$$(4.6\text{-}4) \qquad \sigma_{\epsilon-\delta}(A + D) \subset \sigma_\epsilon(A) \subset \sigma_{\epsilon+\delta}(A + D),$$

$$(4.6\text{-}5) \qquad \sigma_\epsilon(A) \subset W_\epsilon(A).$$

Proof. To demonstrate (4.6-4), given λ an eigenvalue of some $A + E$ with $\|E\| \leq \epsilon$, then λ is an eigenvalue of $A + D + (E - D)$ and $\|E - D\| \leq \epsilon + \delta$. If $\delta > \epsilon$, $\sigma_{\epsilon-\delta}$ is taken to be the null set. As to (4.6-5), from $(A+E)x = \lambda x$ and $\|x\| = 1$, we know that

$$(4.6\text{-}6) \qquad |\langle Ax, x \rangle - \lambda| = \|Ex\| \leq \epsilon. \quad \square$$

In the preceding three sections of this chapter, we showed connections between the numerical analysis methods of finite differences and finite elements and numerical range considerations. The third basic numerical analysis method for partial differential equations is that of finite spectral methods. This method, which relies on the ability to do fast computational integrals or transforms, offers very high resolution in practice but considerable difficulty in other ways, such as boundary values, stability, and proofs of convergence. One motivation for developing a theory of pseudo-eigenvalues was to help with the stability of spectral methods. Thus the considerations of this section may be regarded as extensions and variations on Theorem 4.3-1 by replacing the needed estimates with respect to the unit disk $|z| \leq 1$ by an arbitrary stability region.

Notes and References for Section 4.6

The first definition of ϵ-pseudo-eigenvalues was apparently by
J. M. Varah (1979). "On the Separation of Two Matrices," *SIAM J. Numer. Anal.* **16**, 216–222.
where they were called "ϵ-eigenvalues." Considerable work more recently began in
L. N. Trefethen (1990). "Approximation Theory and Numerical Linear Algebra," in *Algorithms for Approximation II*, eds. J. Mason and M. Cox, Chapman, London, 336–360.
where they were originally called "approximate eigenvalues." An application to numerical methods of spectral type may be found in
S. Reddy and L. N. Trefethen (1992). "Stability and the Method of Lines," *Numer. Math.* **62**, 235–267.
where it is shown that a necessary and sufficient condition for stability is that the ϵ-pseudo-spectrum of the spatial discretization operator lie within a distance $O(\epsilon) + O(k)$ of the stability region as ϵ and the time step k tend to zero.

It has been known for a long time by experienced practitioners in numerical modeling in which the discretizations involve nonnormal operators that one must go beyond the spectrum $\sigma(A)$ to understand stability. See, for example, the discussion in
P. J. Schmid, D. S. Henningson, M. Khorrami, and M. R. Malik (1993). "A Study of Eigenvalue Sensitivity for Hydrodynamic Stability Operators," *Theoret. Comput. Fluid Dynamics* **4**, 227–240.
where the sensitivity properties of the eigenvalues are related to transient properties of the flow. The introduction of the notion of pseudo eigenvalues is a mathematical way of trying to create a general theory for such instabilities, see
L. N. Trefethen, A. E. Trefethen, S. Reddy, T. Driscoll (1993). "Hydrodynamic Stability without Eigenvalues," *Science* **261**, 578–584.

where it is shown that very small perturbations to even very smooth (Couette) flow can produce error amplification of $O(10^5)$ by a linear mechanism, even though all eigenmodes decay monotonically.

An interesting predecessor paper to pseudo eigenvalues is
S. Parter (1962). "Stability, Convergence, and Pseudo-Stability of Finite-Difference Equations for an Over-Determined Problem," *Numerische Mathematik* **4**, 277–292.

As we mentioned in Section 4.5, the Lax–Wendroff scheme is generally stable so long as

$$(|u| + c)\frac{\Delta t}{\Delta x} < 1.$$

The mathematical meaning of this is, roughly, that the scheme should not advance faster than its characteristics. The physical meaning is, roughly, that we want to prevent the simulated Euler equations (4.4–15) from developing shocks. In the paper just mentioned, the easier transport equation

$$u_t + a(x,t)u_x = g$$

is considered along with initial and boundary conditions, and it is shown that for the Lax–Wendroff scheme the Δt for stability must satisfy

$$a\frac{\Delta t}{\Delta x} \leqq 1$$

but that there is an interval of "pseudo-stability"

$$1 < a\frac{\Delta t}{\Delta x} < \sqrt{4/3}$$

in which the scheme is generally unstable and nonconvergent, but nonetheless single errors are eventually damped out.

Clearly (see Theorem 4.3-1), the notion of ϵ-pseudo-spectra is related to resolvent operators possessing certain estimates in terms of the distance of λ to the spectrum. Recall that for any operator T we know by (1.4-6) and Schwarz's inequality that

(4.6-7) $\qquad d(\lambda, W(T)) \leqq \|(T - \lambda)^{-1}\|^{-1} \leqq d(\lambda, \sigma(T)).$

This is the basic framework within which the theory and application of ϵ-pseudo-spectra lie. It is also the framework of several of the classes of operators that we will study in Chapter 6.

We remark that the notion and results for ϵ-pseudo-eigenvalues for matrices could be extended to arbitrary operators, e.g., by replacing the point spectra $\sigma_p(A)$ by the full spectrum $\sigma(T)$ and the numerical range $W(A)$ by its closure $\overline{W(A)}$, but we shall not investigate such generalizations here.

Endnotes for Chapter 4

Certainly the strongest argument for bringing numerical range and numerical analysis closer together in the future is the stability of $W(T)$ under small perturbations. As we stated in the Notes of Section 4.6, it has been long known in scientific circles and the numerical analysis community that once one encounters nonnormal operators or nonnormal discretizations of physical processes, small errors in data or measurement can produce potentially highly erroneous simulations. This lore underlies the recent interest in pseudo eigenvalues or, put another way, the latter is a recent manifestation of the former.

Indeed, it is quite amusing to select an $n \times n$ matrix A at random and numerically plot its numerical range $W(A)$: algorithms to do this will be explained in the next chapter (Section 5.6). Then change an element or two of A slightly and you will see very little movement of $W(A)$. This can be in great contrast to what is happening to the spectrum. Thus, if one can establish convergence and/or stability criteria in terms of $W(A)$ or $w(A)$, rather than in terms of eigenvalues or $r(A)$, one is far more secure against incorrect conclusions due to physical data measurement precision limitations or machine epsilon occurrences. It should also be stressed that even if a physical theory can be given entirely in terms of symmetric or normal operators, actual physical measurements will always carry some departure or truncation from such theory.

Another point has recently been emphasized by
M. Eiermann (1993). "Fields of Values and Iterative Methods," *Linear Algebra Appl.* **180**, 167–197,
namely, that if A is not normal, only asymptotic behavior of an iterative numerical method can be drawn from spectral information. To understand the progress of an iteration after a finite number of steps, one can often better use the numerical range. A similar viewpoint is taken also in the recent papers by
G. Starke (1993). "Fields of Values and the ADI Method for Non-normal Matrices," *Linear Algebra Appl.* **180**, 199–218.
O. Axelsson, H. Lu and B. Polman (1994). "On the Numerical Radius of Matrices and its Application to Iterative Solution Methods," *Linear and Multilinear Algebra* **37**, 225–238.

Let us comment that even though the terminologies "ϵ-eigenvalues" and "approximate eigenvalues" gave way to the term "ϵ-pseudo- eigenvalues," it should be noted that the term "pseudo-eigenvalue" appears in the theory of spectral concentration, where an unstable eigenvalue of a Hamiltonian operator has been converted by a perturbation to a point of steep spectral slope in the continuous spectrum. Thus it could turn out that the original terminology ϵ-eigenvalue and ϵ-spectra were not inappropriate.

The recently discovered fundamental connection between the operator trigonometry of Chapter 3 and iterative methods for the solution of large sparse systems $Ax = b$, goes beyond steepest descent and conjugate gradient algorithms as described in Sections 4.1 and 4.2, respectively. For some of the recent developments, see

K. Gustafson (1996) "Operator Trigonometry of Iterative Methods," *Numer. Lin. Alg. Applic.*, to appear.

K. Gustafson (1996). "Operator Trigonometry of the Model Problem," to appear.

For example, the classic Richardson iteration $x_{k+1} = x_k + \alpha(b - Ax_k)$ with iteration matrix $G_\alpha = I - \alpha A$ is optimized at convergence rate $\rho_{\text{opt}} = \sin A$. As another example, Chebyshev polynomial preconditionings may be seen to be optimized at midwidth $(A)/\text{center}(A) = \sin A$. Similarly, Jacobi, SOR, and SSOR achieve optimal convergence rates described by $\sin \tilde{A}, \sin^{1/2} \tilde{A}^{1/2}$, and $\sin \tilde{A}^{1/2}$, respectively, where the operators $\tilde{A}$ are defined in terms of upper and lower bounds for A relative to A's diagonal or upper or lower triangular parts.

Moreover, the observed superlinear convergence rate of conjugate gradient methods can be explained as follows. The $\sin(A^{1/2})$ convergence rate of (4.2-1) can be expressed as

$$\sin(A^{1/2}) = \frac{\lambda_{\max}^{1/2} - \lambda_{\min}^{1/2}}{\lambda_{\max}^{1/2} + \lambda_{\min}^{1/2}} \sim 1 - 2\left(\frac{\lambda_1}{\lambda_n}\right)^{1/2}.$$

In gradient descent the error converges to the subspace $V = \text{sp}\{x_1, x_n\}$. However, the conjugate gradient error moves toward $V^\perp$. On $V^\perp$ the operator A has reduced spectrum $\lambda_2 \leq \cdots \leq \lambda_{n-1}$. Thus the conjugate gradient error rate improves to

$$\sin((A|_{V^\perp})^{1/2}) = \frac{\lambda_{n-1}^{1/2} - \lambda_2^{1/2}}{\lambda_{n-1}^{1/2} + \lambda_2^{1/2}} \sim 1 - 2\left(\frac{\lambda_2}{\lambda_{n-1}^{1/2}}\right).$$

In computational partial differential equations, a key model problem for testing iterative solvers is the Dirichlet Problem on the unit square. Moreover, often in applications it is precisely this computation which consumes the bulk of the computing time. For expositions of the properties of important iterative solvers on this model problem, see the example

D. M. Young (1971). *Iterative Solution of Large Linear Systems*, Academic Press, New York.

W. Hackbusch (1994). *Iterative Solution of Large Sparse Systems of Equations*, Springer-Verlag, Berlin.

For this model problem, with discretized Laplacian A_h and finite difference grid size h, it is now known that $\sin A_h = \cos(\pi h)$ and that the first antieigenvalue of A_h is $\mu = \sin(\pi h)$. Thus the maximum turning angle $\phi(A_h)$ increases with decreasing grid size h. Moreover, the full operator trigonometry of A_h may be seen to depend upon a set of related "harmonic" grids.

5

Finite Dimensions

Introduction

The theory of numerical range in finite-dimensional spaces is very rich and varied. In fact, a lot of recent research has been focused on the numerical range, and its variations, in finite dimensions. Avoiding the evidently impossible task of doing justice to all of the work done in this field, we attempt to present a representative selection and hope that it covers all the basic material.

Consistent with the rest of the book, we use the usual inner product in $\mathbb{C}^n$. Thus, we have for $x, y \in \mathbb{C}^n$, $x = (x_1, x_2, \ldots, x_n)$, $y = (y_1, y_2, \ldots, y_n)$,

$$\langle x, y \rangle = x_1 \bar{y}_1 + x_2 \bar{y}_2 + \cdots + x_n \bar{y}_n.$$

If A is an operator represented by an $n \times n$ matrix $A = (a_{ij})$, then $\|A\| = \sup\{\|Ax\|, x \in \mathbb{C}^n$, $n \times n$ matrices acting in $\mathbb{C}^n$.

In this chapter, we study some basic properties of the field of values and its localization. Further, we look at two special topics that have received considerable attention, namely Hadamard products and special (not classical) numerical ranges. We also indicate how one may compute $W(A)$ for matrices A and give some examples.

5.1. Value Field

We already know that the numerical range is convex (Theorem 1.1-2). In addition, we have the following result.

Theorem 5.1-1. *The numerical range of any matrix A acting in $\mathbb{C}^n$ is compact and the numerical radius attained.*

Proof. The function $x \to \langle Tx, x \rangle$ from the compact set $S = \{x : \|x\| = 1\}$ into $\mathbb{C}$ is continuous. Further, the real-valued function $x \to |\langle Tx, x \rangle|$ from S to R attains its maximum value. $\square$

We note that the spectral inclusion is more transparent in finite dimensions since for any $\lambda \in \sigma(T)$, we have $\lambda x = Tx$, $\|x\| = 1$, and $\langle \lambda x, x \rangle = \langle Tx, x \rangle$. Furthermore, we have the following inclusion theorem for the numerical ranges of submatrices.

Theorem 5.1-2 (Submatrix inclusion). *Let $A = (a_{ij}) \in M_n$. Let $A(J)$ denote the submatrix having the elements of A in the rows and columns given by an index set $J \subset \{1, 2, \ldots, n\}$. Then $W(A(J)) \subset W(A)$.*

Proof. Let $J = \{j_1, j_2, \ldots, j_k\}$, $1 \leq j_1 \leq j_2 \leq \cdots \leq j_k \leq n$. Let $x \in \mathbb{C}^k$, $\|x\| = 1$. By inserting zeros in the right places, we can obtain a vector $y \in \mathbb{C}^n$ such that $\|y\| = 1$. The vector y looks like $(\ldots 0, x_{j_1}, 0, \ldots, x_{j_2}, 0, \ldots, x_{j_k}, 0, \ldots)$. Then $\langle T(J)x, x \rangle = \langle Ty, y \rangle \in W(T)$. □

In the case of a symmetric matrix, the special inclusion has an additional feature.

Theorem 5.1-3. *The numerical range of a symmetric matrix A is the real interval $[m, M]$, where m and M are the least and greatest eigenvalues of A, respectively.*

Proof. We know that $W(A)$ is a compact convex set on the real line. Let $W(A) = [m, M]$. Since $W(A)$ is closed, $m, M \in \sigma(A)$ by Theorem 1.2-4. By the spectral inclusion, we conclude that m and M are the minimum and maximum values of $\sigma(A)$. □

Theorem 5.1-4. *The numerical range of a unitary matrix A is a polygon inscribed in the unit circle.*

Proof. Let $Ax = \lambda x$, $\|Ax\|^2 = \|x\|^2 = |\lambda|^2 \|x\|^2$. Thus λ is on the unit circle. Since A is normal, $W(A) = \operatorname{co} \sigma(A)$ by Theorem 1.4-4, and we have $W(A) = \operatorname{co}(\lambda_1, \lambda_2, \ldots, \lambda_n)$, $|\lambda_i| = 1$, $i = 1, 2, \ldots, n$. □

Let us now look at some applications of the numerical range in matrix theory. In particular, we prove a Schur triangularization and a spectral decomposition of a normal matrix.

Theorem 5.1-5. *For each $T \in M_2$, there is a unitary matrix $U \in M_2$ such that the two main diagonal entries of U^*TU are equal.*

Proof. Consider $B = T - \frac{1}{2}(\operatorname{tr} A)I$. Then $\operatorname{tr} B = 0 = $ the sum of the eigenvalues of B. If λ and $-\lambda$ are the eigenvalues of B, we have $\lambda \in W(B)$ and $-\lambda \in W(B)$ by spectral inclusion, $\frac{\lambda}{2} - \frac{\lambda}{2} = 0$ is a convex combination of two elements of $W(B)$, and hence $0 \in W(B)$. So, there is a vector $v \in \mathbb{C}^2$, $v = (v_1, v_2)$, $\|v\| = 1$, such that $\langle Bv, v \rangle = 0$.

Define the unitary operator U by the matrix

$$\begin{bmatrix} v_1 & -\bar{v}_2 \\ v_2 & \bar{v}_1 \end{bmatrix}.$$

Then, we can see that

$$U^*BU = \begin{bmatrix} 0 & a \\ b & c \end{bmatrix},$$

$a, b, c, \in \mathbb{C}^2$, and $c = 0$ since $\operatorname{tr} B = 0$. □

Theorem 5.1-6. *For each $T \in M_n$, there is a unitary matrix $U_n \in M_n$ such that all the main diagonal elements of U^*TU are equal.*

Proof. We use induction with Theorem 5.1-5 as the starting point. Let $B = T - \frac{1}{n}(\operatorname{tr} T)I$. Then $\operatorname{tr} B = 0$. If $\lambda_1, \lambda_2, \ldots, \lambda_n$ are the eigenvalues (possibly repeated) of B, we have $\sum_1^n \lambda_k = \operatorname{tr} B = 0$. Since for each $k = 1, 2, \ldots, n$, $\lambda_k \in W(B)$, we have $\frac{1}{n}\lambda_1 + \cdots + \frac{1}{n}\lambda_n = 0 \in W(B)$ by the convexity of $W(B)$. Hence, there is a unit vector $v \in \mathbb{C}^n$ such that $\langle Bv, v \rangle = 0$. Let V_n be a unitary matrix with v as the first column. Then V_n^*BV has the matrix

$$\begin{bmatrix} 0 & a \\ b & D_{n-1} \end{bmatrix},$$

where $a = (a_2, \ldots, a_n)$, $b = (b_2, b_3, \ldots, b_n)^*$, $a, b \in \mathbb{C}^{n-1}$, and $D_{n-1} \in M_{n-1}$.

By the induction hypothesis, there exists a unitary matrix $V_{n-1} \in M_{n-1}$ such that $V_{n-1}^*D_{n-1}V_{n-1}$ has all the main diagonal elements equal. Extend the unitary matrix V_{n-1} to a unitary matrix W_n by

$$W_n = \begin{bmatrix} 1 & 0 \\ 0 & V_{n-1} \end{bmatrix}.$$

Then $U_n = V_nW_n$ is the desired unitary matrix. □

We will now use the numerical range to develop a sequence of results leading to the spectral decomposition of a normal operator.

Let us first recall that (Theorems 1.4-1 to 1.4-5) when the matrix A is normal, $r(A) = w(A) = \|A\|$, $W(A) = \operatorname{co} \sigma(A)$, and the extreme points are eigenvalues if $W(A)$ is closed. We then immediately have the following theorem.

Theorem 5.1-7. *The extreme points of $W(A)$ are eigenvalues of a normal matrix A.*

Proof. $W(A)$ is closed (Theorem 5.1-1). Now use Theorem 1.4-5. □

The boundary of the numerical range of a normal A has special relations to its spectrum, as the following theorems reveal.

Theorem 5.1-8. *Let A be a normal matrix. Then*

(5.1-1) $$\partial W(A) \cap \sigma(A) \neq \emptyset.$$

Proof. $W(A)$ is a compact convex set and hence is the convex hull of its extreme points (a nonempty set). These extreme points are eigenvalues (Theorem 5.1-7). □

The last result, apparently trivial, leads us to the spectral decomposition of a normal matrix. We start with an eigenvalue on the boundary and the corresponding eigenspace and then proceed to repeat the procedure on the complement. The following theorem reveals the significance of such an eigenvalue.

Theorem 5.1-9. *Let A be a matrix, not necessarily normal, and α an eigenvalue on the boundary of the numerical range. Then, the dimension of the eigenspace M for α is equal to its algebraic multiplicity, and A is unitarily equivalent to $\alpha I_M \oplus B$, where $\alpha \notin \sigma(B)$.*

Proof. Let x be an eigenvector corresponding to α, $\alpha x = Ax$. We can, by use of a unitary transformation, choose x as $(1,0,0,\dots)$ and $Ax = \alpha x = (\alpha, 0, 0, \dots)$. Evidently,

(5.1-2) $$A = \begin{bmatrix} \alpha & a \\ 0 & A_1 \end{bmatrix}.$$

Now let us choose a principal 2×2 submatrix B of A, given by the indices 1 and i, $i \in [2, \dots, n]$,

$$B = \begin{bmatrix} \alpha & c \\ 0 & \beta \end{bmatrix}.$$

B has eigenvalues α and β, $\alpha \in W(B)$ by spectral inclusion, and $W(B) \subset W(A)$ by Theorem 5.1-2. Since $\alpha \in \partial W(B)$, $W(B)$ is an ellipse and α is a focus on the boundary of the ellipse, we must have $W(B) = [\alpha, \beta]$. This implies by Theorem 1.4-1 that B is normal. The condition $B^*B = BB^*$ now implies $c = 0$. Thus, we see that

$$A = \begin{bmatrix} \alpha & 0 \\ 0 & A_1 \end{bmatrix}.$$

If α has multiplicity greater than 1, we can repeat the process until we obtain a unitarily equivalent matrix

$$\begin{bmatrix} \alpha I_k & 0 \\ 0 & A_{k+1} \end{bmatrix},$$

where k is the multiplicity of α. Notice that the k-dimensional eigenspace corresponding to α is a reducing subspace of A. □

Now, let us assume that A is normal. Then A_{k+1} is normal on an $(n-k)$-dimensional space, and for any $b \in \sigma(A)$, $b \neq \alpha$, we have $b \in \sigma(A_{k+1})$. We now use Theorem 5.1-8 and conclude that $b \in \partial W(A_{k+1})$. We can now repeat the argument in Theorem 5.1-9 and obtain the following Corollary.

Corollary 5.1-10. *If the operator A is normal, it can be decomposed in the form*
$$A = A_1 \oplus A_2 \oplus \cdots \oplus A_\ell,$$
where each A_i is normal and $\sigma(A_i) \subset \partial W(A_i)$ for $i = 1, 2, \ldots, \ell$.

Proof. Similar to that of Theorem 5.1-9. □

Corollary 5.1-11. *If $Ax = \alpha x$, $Ay = \lambda y$, $\alpha \neq \lambda$, and $\alpha \in \partial W(A)$, then $\langle x, y \rangle = 0$.*

Proof. From (5.1-1) we have $\lambda \in \sigma(A_1)$ and so the first component of λ is zero. So $\langle x, y \rangle = 0$. □

The last fact can be used to deduce the converse of Corollary 5.1-10.

Theorem 5.1-12. *If A is the direct sum $A = A_1 \oplus A_2 \oplus \cdots \oplus A_\ell$, where each A_i has the property $\sigma(A_i) \subset \partial W(A_i)$, then A is normal.*

Proof. Each A_i is normal, because each A_i has a complete set of orthogonal eigenvectors, by Corollary 5.1-11. Hence A is normal. □

Notes and References for Section 5.1

Most of the elementary properties of $W(A)$ have already been studied in Chapter 1 and apply in the finite-dimensional case. The eigenvalues appearing in Theorem 5.1-9 and the Corollary 5.1-10 appear in
C. R. Johnson (1976). "Normality and the Numerical Range," *Lin. Alg. Appl.* **15**, 89–94.
Such eigenvalues are called normal eigenvalues.

5.2 Gersgorin Sets

The numerical range of a sum of matrices can be located, at least roughly, since
$$W(A+B) \subset W(A) + W(B) = \{\lambda + \mu, \lambda \in W(A), \mu \in W(B)\}.$$
Such a relation is not true for the spectra. So, one way to localize $\sigma(A+B)$ is the inclusion $\sigma(A+B) \subset W(A+B) \subset W(A) + W(B)$. This is perhaps

quite an important motivation in obtaining sets containing $W(T)$. Such set containment was first obtained by Gersgorin (see Notes) for the spectrum. Since the results we mention use proofs similar to those of Gersgorin, it is convenient to look at his proof.

Theorem 5.2-1. Let $A = [a_{ij}] \in M_n$. Define $R_i = \sum_{j \neq i} |a_{ij}|$ and $D_i(A) = \{z : |z - a_{ii}| \leq R_i\}$, $1 \leq i \leq n$. Then $\sigma(A) \subset \bigcup_1^n D_i(A) = D(A)$.

Proof. Let $\lambda \in \sigma(A)$, $Ax = \lambda x$, $x \neq 0$. If x is the n-tuple $(x_1, x_2, \ldots, x_n)$, choose k so that $|x_k| = \max\{|x_1|, |x_2|, \ldots, |x_n|\}$. We then have $\lambda x_k = \sum_{i=1}^n a_{ki} x_i$ and so

(5.2-1) $$\lambda x_k - a_{kk} x_k = \sum_{i \neq k} a_{ki} x_i.$$

Hence $|\lambda - a_{kk}||x_k| \leq \sum_{i \neq k} |a_{ki}||x_i| \leq \sum_{i \neq k} |a_{ki}||x_k|$. Thus $|\lambda - a_{kk}| \leq D_i(A)$.

Observe that we can repeat the above process with the columns of A instead of the rows and obtain

(5.2-2) $$\sigma(A) \subset \bigcup_1^n D_i'(A) = D'(A),$$

where

$$D_i'(A) = \sum_{i \neq j} |A_{ij}|.$$

Thus $\sigma(A) \subset D_i(A) \cap D_i'(A)$. Notice that $D_i'(A) = D_i(A^*)$. $\square$

The following theorem gives a Gersgorin-type set containing the numerical range.

Theorem 5.2-2. Let $A = [a_{ij}] \in M_n$. With the notation of the previous theorem, $W(A)$ is contained in the set

$$S = \text{co}\left[\bigcup_1^n \{z : 2|z - a_{ii}| \leq D_i(A) + D_i'(A)\}\right].$$

Proof. Consider any line of support of the convex set S. We will show that $W(A)$ is also on the same side of this line of support. Without loss of generality, we may assume that S is contained in the right half-plane $\operatorname{Re} z \geq 0$ and show that $W(A)$ is contained therein. By Theorem 5.2-1 applied to $\operatorname{Re} A = \frac{A + A^*}{2} = B$, we have

(5.2-3) $$\sigma(B) \subset \text{co}\left[\bigcup_1^n \{z : 2|z - a_{ii}| \leq D_i(A + A^*)\}\right].$$

Notice that, by the triangle inequality,

(5.2-4) $$D_i(B) \leq \frac{D_i(A)}{2} + \frac{D_i(A^*)}{2}.$$

Since $S \subset \{z : \operatorname{Re} z \geq 0\}$, we have $\operatorname{Re} a_{ii} = b_{ii} > \frac{D_i(A)}{2} + \frac{D'_i(A)}{2}$ and since the last set $\frac{D_i(A)}{2} + \frac{D'_i(A)}{2}$ is contained in $\{z : \operatorname{Re} z \geq 0\}$, so is $D_i(B)$. Hence $\sigma(B) \subset \{z : \operatorname{Re} z \geq 0\}$. As B is selfadjoint, $W(B) = \operatorname{co} \sigma(B)$ and hence $W(B) \subset \{z : \operatorname{Re} z \geq 0\}$.

Since $A = B + \frac{A-A^*}{2}$, we have $W(A) \subset W(B) + W\frac{(A-A^*)}{2}$ and $W\frac{(A-A^*)}{2}$ is purely imaginary. Hence $W(A)$ is contained in the set $\{z : \operatorname{Re} z \geq 0\}$. □

Corollary 5.2-3. *The numerical radius satisfies*
(5.2-5)
$$w(A) \leq \max_i \left(a_{ii} + \frac{D_i(A)}{2} + \frac{D'_i(A)}{2} \right) = \max_i \left(\sum_{j=1}^n \frac{|a_{ij}| + |a_{ji}|}{2} \right).$$

Proof. The right-hand side of this inequality is the maximum absolute value of the elements in S. □

Another method of obtaining a Gersgorin set containing $W(A)$ makes use of the following interesting fact. If we subtract the diagonal from any matrix, we have a matrix with zero trace. Thus, if $R = [r_{ij}]$ and

$$R^0 = R - \begin{bmatrix} r_{11} & & & \\ & r_{22} & & \\ & & \ddots & \\ & & & r_{nn} \end{bmatrix},$$

the sum of the eigenvalues of R^0 is zero. If, in addition, R^0 is selfadjoint, we can conclude that some of the eigenvalues are negative and the others positive (unless all of them are zeros). Then, the eigenvalues of R^0 lie between R^0_m and R^0_M, the minimum and maximum eigenvalues of R^0. The following theorem applies this observation to the real and imaginary parts of any given matrix.

Theorem 5.2-4. *Let $A = [a_{ij}] \in M_n$, $A = R + iS$. Then*

$$W(A) \subset \operatorname{co} \left[\bigcup_1^n B_i \right],$$

where
(5.2-6) $\quad B_i = \{z : R^0_m \leq \operatorname{Re}(z - a_{ii}) \leq R^0_M, S^0_m \leq \operatorname{Im}(z - a_{ii}) \leq S^0_M\}.$

Proof. Let $f \in \mathbb{C}^n$, $f = (f_1, f_2, \ldots, f_n)$, $\|f\| = 1$. Then $y = \langle Af, f \rangle - \sum_1^n a_{ii}|f_i|^2$, we have $\operatorname{Re} y = \langle R^0 f, f \rangle$, and $\operatorname{Im} y = \langle S^0 f, f \rangle$. Obviously,

116   5. Finite Dimensions

$R_m^0 \leq \operatorname{Re} y \leq R_M^0$ and $S_m^0 \leq \operatorname{Im} y \leq S_M^0$. Since $\langle Af, f \rangle$ differs from y by some element of the convex hull of the a_{ii}, the result follows. □

Corollary 5.2-5. $W(A) \subset \{z : |\operatorname{Re} z| \leq P, |\operatorname{Im} z| \leq \sigma\}$, where $P = \max_i \sum_j |r_{ij}^0|$ and $\sigma = \max_i \sum_j |s_{ij}^0|$.

Proof. We have, using Theorem 5.2-1 and the notation of Theorem 5.2-4,

$$|R_m^0| \leq P \quad \text{and} \quad |R_m^0| \leq P, \quad |S_m^0| \leq \sigma, \quad |S_M^0| \leq \sigma. \quad \square$$

Example 1. Let us first look at a perturbation of a shift in which all the diagonal elements are zeros:

$$A = \begin{bmatrix} 0 & 1 & 0 & 0 \\ 0 & 0 & 1 & 0 \\ 0 & 0 & 0 & 1 \\ \frac{1}{2} & 0 & 0 & 0 \end{bmatrix}.$$

By Theorem 5.2-2,

$$D_1(A) = D_2(A) = D_3(A) = 1 \quad \text{and} \quad D_4(A) = \frac{1}{2},$$

$$D_1'(A) = \frac{1}{2}, \quad D_2'(A) = D_3'(A) = D_4'(A) = 1,$$

$$W(A) \subset \{z : |z| \leq 1\}$$

On the other hand, the minimum and maximum eigenvalues of R^0 and S^0 are, respectively, $(-0.89, 0.89)$ and $(-0.89, 0.89)$. Thus by Theorem 5.2-4, $W(A)$ is contained in the rectangle

$$\{-0.89 \leq \operatorname{Re} z \leq 0.89, \ -0.89 \leq \operatorname{Im} z \leq 0.89\}.$$

The actual numerical range is shown in Fig. 1 of Section 5.6.

Example 2. When the diagonal elements are not all equal, we get a union of circles to bound $W(A)$. Consider

$$A = \begin{bmatrix} 1 & 2 & 0 \\ 0 & 2 & 2 \\ 0 & 0 & 3 \end{bmatrix}.$$

By Theorem 5.2-2,

$$D_1(A) = D_2(A) = 2, \quad D_3(A) = 0,$$

$$D_1'(A) = 0, \quad D_2'(A) = D_3'(A) = 2.$$

Thus $W(A)$ is contained in the convex hull of the union of the circles $\{z : |z - 1| \leq 1\}$, $\{z : |z - 2| \leq 2\}$, and $\{z : |z - 3| \leq 1\}$. On the other hand, using Theorem 5.2-4, $W(A)$ is contained in the convex hull of the three rectangles, $R_1, R_2,$ and R_3:

$$R_1 = \{z : -1.414 \leq \mathrm{Re}\,(z-1) \leq 1.414;\ -1.414 \leq \mathrm{Im}\,(z-1) \leq 1.414\}$$
$$R_2 = \{z : -1.414 \leq \mathrm{Re}\,(z-2) \leq 1.414;\ -1.414 \leq \mathrm{Im}\,(z-2) \leq 1.414\}$$

and

$$R_3 = \{z : -1.414 \leq \mathrm{Re}\,(z-3) \leq -1.414;\ -1.414 \leq \mathrm{Im}\,(z-3) \leq 1.414\}.$$

Figure 2 in Section 5.6 shows $W(A)$ in this case.

Example 3. Let us now look at the companion matrix of the polynomial $\lambda^3 - 3\lambda^2 + 4\lambda - 2$ (see the next section),

$$A = \begin{bmatrix} 0 & 1 & 0 \\ 0 & 0 & 1 \\ 2 & -4 & 3 \end{bmatrix}.$$

By Theorem 5.2-2,

$$D_1(A) = 1, \quad D_2(A) = 1, \quad D_3(A) = 6,$$

$$D'_1(A) = 2, \quad D'_2(A) = 5, \quad D'_3(A) = 1,$$

and hence $W(A) \subset \mathrm{co}\,\{C_1 \cup C_2 \cup C_3\}$, where $C_1 = \{z : |z| \leq \frac{3}{2}\}$, $C_2 = \{z : |z| \leq 3\}$ and $C_3 = \{z : |z-3| \leq \frac{7}{2}\}$. Theorem 5.2-4 gives $W(A) \subset \mathrm{co}\,\{R_1, R_2\}$, where

$$R_1 = \{z : -2.06 \leq \mathrm{Re}\,z \leq 4.57;\ -2.74 \leq \mathrm{Im}\,z \leq 2.74\},$$
$$R_2 = \{z : -2.06 \leq \mathrm{Re}\,(z-3) \leq 4.57;\ -2.74 \leq \mathrm{Im}\,z \leq 2.74\}.$$

$W(A)$ is given in Fig. 3 of Section 5.6.

Notes and References for Section 5.2

The original Gersgorin result was given in
S. Gersgorin (1931). "Über die Abrenzung der Eigenwerte einer Matrix," *Izv. Akad. Nauk SSSR* (Ser. Mat.) 7, 749–754.

An early paper connecting the numerical range to the Gersgorin theory was
F. Bauer (1968). "Fields of Values and Gershgorin Disks," *Numerische Math.* **12**, 91–95.

A Gersgorin set for the numerical range in Theorem 5.2-2 was first given by
C. R. Johnson (1973). "A Gersgorin inclusion set for the field of values of a finite matrix," *Proc. Amer. Math. Soc.* **41**, 57–60.
It provides a fair approximation and makes equal use of the rows and columns.

For any operator $T = R + iS$, we have

$$W(T) \subset W(R) + iW(S).$$

Thus, we always have the result that $W(T)$ is contained in the rectangle determined by the minimum and maximum eigenvalues of R and S. This result is known as the *Bendixson–Hirsch theorem*.

Theorem 5.2-4 is a modification of this theorem obtained by deleting the diagonal. Further details can be seen in

A. I. Mees and D. P. Atherton (1979). "Domains Containing the Field of Values of a Matrix," *Lin. Alg. Appl.* **26**, 289–296.

Other Gersgorin sets for the spectrum and the numerical range can be found in

V. N. Solov'ev (1983). "A Generalization of Gersgorin's Theorem," *Izv. Akad. Nauk. SSSR Ser. Mat.* **47**, 1285–1302; English translation in *Math. USSR Izv.* **23** (1984),

A. A. Abdurakmanov (1988). "The Geometry of the Hausdorff Domain in Localization Problems for the Spectrum of Arbitrary Matrices," *Math. USSR Sbornik.* **59**, 39–51.

An excellent and easy-to-read account of Gersgorin sets for the spectrum is given in

R. A. Brualdi and S. Mellendorf (1993). "Regions in the Complex Plane Containing the Eigenvalues of a Matrix," *Amer. Math. Monthly* **101**, 975–985.

5.3 Radius Estimates

Upper bounds on the numerical radius, although elusive in the general case, are more available in some special cases. We consider, in particular, the cases of 0–1 matrices and companion matrices.

A square matrix all of whose elements are 0 or 1 is called a *0–1 matrix*.

Example.

$$A = \begin{bmatrix} 0 & 0 & 0 & 1 & 0 & 0 \\ 0 & 0 & 0 & 0 & 0 & 0 \\ 1 & 0 & 0 & 0 & 0 & 0 \\ 0 & 0 & 1 & 0 & 0 & 0 \\ 0 & 1 & 0 & 0 & 0 & 0 \\ 0 & 0 & 0 & 0 & 1 & 0 \end{bmatrix}.$$

Let us consider the general case in which A is a 0–1 matrix with at most one 1 in each row and each column. We can associate, in a unique fashion, an injection σ, $\sigma : x \to \{1, 2, \ldots, n\}$, where $x \subset \{1, 2, \ldots, n\}$ and $A = A(\sigma)$, where $A(\sigma)_{i,j} = x(j \in x)\delta_{i,\sigma(j)}$. So, the i, j component of $A = A(\sigma)$ is 1 only when $j \in x$ and $\sigma(j) = 1$. Otherwise, it is zero.

5.3 Radius Estimates

Let us, before going on, illustrate the injection σ in the case of the matrix A given above. In column one, the 1 appears in row 3, so $\sigma(1) = 3$, and in column three the 1 appears in column 4, so $\sigma(3) = 4$ and $\sigma(4) = 1$. Thus, we complete a cycle (1,3,4). Similarly, $\sigma(2) = 5$, $\sigma(5) = 6$, and there is no 1 in column 6. Also there is no 1 in the second row. Thus, we terminate a chain (open circuit) (2,5,6] of length 3. Notice that the cycle (1,3,4) and the chain are disjoint.

Returning to our general discussion, we can thus write σ as the composition $\theta * \tau$, where θ is the cycle and τ the chain. Thus $\sigma = \theta * \tau$, $A(\sigma) = A$. If $\phi \in S_n$ the symmetric group of n, we can show that $A(\phi\sigma) = A(\phi)A(\sigma)$ and $A(\sigma\phi) = A(\sigma)A(\phi)$. In fact, $A(\phi\sigma) = x(j \in x, \sigma(j) = k, \sigma(k) = i) = x(j \in x$ and $\phi\sigma(j) = i) = A(\phi\psi)_{i,j}$, where

$$\sigma = \sigma_1 * \sigma_2 * \cdots * \sigma_k * \theta_1 * \cdots * \theta_\ell,$$

where the σs are chains, observing that there may be degenerate cases (singletons). Let α_i, β_i denote the lengths of σ_i and θ_i and

$$\gamma_t = \sum_{i=1}^{t} \alpha_i, \quad 1 \leq t \leq k, \quad \gamma(0) = 0,$$

and $\gamma_t = \gamma_k + \sum_{i=1}^{t-\beta} \beta_i$, $p+1 \leq t \leq n$. It can be shown (see Marcus and Shure (1979), Notes and References at the end of this section) that there is a $\phi \in S_n$ such that simultaneously

$$\phi\sigma_i\phi^{-1} = \tilde{\sigma}_i, \quad i = 1, 2, \ldots, k, \quad \phi(\theta_j) = \tilde{\theta}_j, \quad i = 1, 2, \ldots, \ell,$$

where $\tilde{\sigma}_t = (\gamma_{t-1}+1, \gamma_{t-1}+2, \ldots, \gamma_t)$, $t = 1, 2, \ldots, k$, and $\tilde{\theta}_t = (\gamma_{p+t-1}+1, \ldots, r_{p+s}]$, $1 \leq s \leq \ell$. Notice that the cumbersome notation is needed to have the elements of successive cycles and chains from left to right in the order $1, 2, \ldots, n$.

We can now in principle calculate $W(A)$ by writing A as a direct sum operator using two kinds of matrices. Let P_m be the $m \times m$ permutation matrix that has 1 in the subdiagonal and 1 in the top right-hand corner in the position $(1, m)$

$$P_m = \begin{bmatrix} 0 & 0 & & & 1 \\ 1 & 0 & 0 & & \\ 0 & 1 & 0 & & \\ \vdots & & & \ddots & \\ 0 & 0 & & 1 & 0 \end{bmatrix}.$$

Let us observe that $w(P_m) = 1$, because we may obtain

$$\sup\{|x_m x_1 + x_1 x_2 + \cdots + x_{m-1} x_m|, \ |x_1|^2 + \cdots + |x_m|^2 = 1\} = 1$$

5. Finite Dimensions

by taking $|x_i| = \frac{1}{\sqrt{m}}$, $i = 1, 2, \ldots, m$. Also, $\|P_m\| = 1$. Let S_m denote the m-shift

$$\begin{bmatrix} 0 & 0 & \cdots & & \vdots \\ 1 & 0 & 0 & & \\ 0 & 1 & 0 & & \\ 0 & & & & 0 \\ 0 & 0 & \cdots & 1 & 0 \end{bmatrix}.$$

Recall that $w(S_m) = \cos\frac{\pi}{m+1}$ (Section 1.3). Notice that the $\tilde{\sigma}_t$ and $\tilde{\theta}_t$ correspond to P_m and Q_m. We can now write

$$(5.3\text{-}1) \qquad A = P_{\alpha_1} \oplus P_{\alpha_2} \oplus \cdots \oplus P_{\alpha_k} \oplus Q_{\beta_1} \oplus \cdots \oplus Q_{\beta_\ell}.$$

In principle, $W(A)$ can now be calculated as the convex hull of this direct sum.

Notice that $w(P_i) = 1$ for all $i \in \{\alpha_1, \ldots, \alpha_k\}$ and $w(Q_i) = \cos\frac{\pi}{i+1}$, $i \in \{\beta_1, \ldots, \beta_\ell\}$. To summarize, we may state the following theorem.

Theorem 5.3-1 (Graph radius). *Let A be a square $n \times n$ 0–1 matrix with at most one 1 in each row and column. Then $w(A) = 1$ if the incidence graph $G(A)$ possesses a nontrivial cycle; otherwise $w(A) = \cos(\pi/(\beta+1))$, where β is the number of vertices in the longest chain in the graph.*

Proof. This follows from the preceding discussion. Observe that the above computation does not necessarily depend on every row or column having at most one 1. The essential condition is that σ can be broken into cycles and chains. □

Figure 4 of Section 5.6 shows $W(A)$ for the 6×6 0–1 matrix A given above. Let us consider some other examples.

Example. Let S be the nine-dimensional right shift S_9

$$S = \begin{bmatrix} 0 & 0 & 0 & 0 & 0 & 0 & 0 & 0 & 0 \\ 1 & 0 & 0 & 0 & 0 & 0 & 0 & 0 & 0 \\ 0 & 1 & 0 & 0 & 0 & 0 & 0 & 0 & 0 \\ 0 & 0 & 1 & 0 & 0 & 0 & 0 & 0 & 0 \\ 0 & 0 & 0 & 1 & 0 & 0 & 0 & 0 & 0 \\ 0 & 0 & 0 & 0 & 1 & 0 & 0 & 0 & 0 \\ 0 & 0 & 0 & 0 & 0 & 1 & 0 & 0 & 0 \\ 0 & 0 & 0 & 0 & 0 & 0 & 1 & 0 & 0 \\ 0 & 0 & 0 & 0 & 0 & 0 & 0 & 1 & 0 \end{bmatrix},$$

and

$$T = S^3 + S^7 = \begin{bmatrix} 0 & 0 & 0 & 0 & 0 & 0 & 0 & 0 & 0 \\ 0 & 0 & 0 & 0 & 0 & 0 & 0 & 0 & 0 \\ 0 & 0 & 0 & 0 & 0 & 0 & 0 & 0 & 0 \\ 1 & 0 & 0 & 0 & 0 & 0 & 0 & 0 & 0 \\ 0 & 1 & 0 & 0 & 0 & 0 & 0 & 0 & 0 \\ 0 & 0 & 1 & 0 & 0 & 0 & 0 & 0 & 0 \\ 0 & 0 & 0 & 1 & 0 & 0 & 0 & 0 & 0 \\ 1 & 0 & 0 & 0 & 0 & 0 & 0 & 0 & 0 \\ 0 & 1 & 0 & 0 & 0 & 1 & 0 & 0 & 0 \end{bmatrix}.$$

We discussed these operators in Section 2.5 (actually, their adjoints, but it makes no difference). If $x = (x_1, x_2, \ldots, x_9) \in \mathbb{C}^9$, we have

(5.3-2) $\quad \langle Tx, x \rangle = x_1 x_4 + x_2 x_5 + x_3 x_6 + x_4 x_7 + x_5 x_8 + x_6 x_9 + x_2 x_9.$

To make explicit the role of the permutations mentioned earlier, in calculating the numerical radius, let us calculate $w(T)$ by using a permutation. Let us write $y = (y_1, y_2, \ldots, y_9)$, $x = (x_1, x_2, \ldots, x_9) = (y_7, y_4, y_1, y_8, y_5, y_2, y_9, y_6, y_3)$, where the permutation is obvious. Then $\langle Tx, x \rangle = \sum_{i=1}^{8} y_i y_{i+1}$. Hence $w(T) = \cos \frac{\pi}{10}$ (see Section 1.3).

Figure 5 of Section 5.6 shows the full numerical range $W(T)$ for this 9×9 matrix T.

Let us now form the matrix product

$$ST = S^4 + S^8 = \begin{bmatrix} 0 & 0 & 0 & 0 & 0 & 0 & 0 & 0 & 0 \\ 0 & 0 & 0 & 0 & 0 & 0 & 0 & 0 & 0 \\ 0 & 0 & 0 & 0 & 0 & 0 & 0 & 0 & 0 \\ 0 & 0 & 0 & 0 & 0 & 0 & 0 & 0 & 0 \\ 1 & 0 & 0 & 0 & 0 & 0 & 0 & 0 & 0 \\ 0 & 1 & 0 & 0 & 0 & 0 & 0 & 0 & 0 \\ 0 & 0 & 1 & 0 & 0 & 0 & 0 & 0 & 0 \\ 0 & 0 & 0 & 1 & 0 & 0 & 0 & 0 & 0 \\ 1 & 0 & 0 & 0 & 1 & 0 & 0 & 0 & 0 \end{bmatrix} = TS.$$

We have

(5.3-3) $\quad \langle STx, x \rangle = \underbrace{x_1 x_5 + x_5 x_9 + x_9 x_1}_{\text{cycle}} + x_2 x_6 + x_3 x_7 + x_4 x_8.$

We can immediately conclude that $w(ST) = 1$ because of the presence of a cycle. Notice that choosing $x_1 = x_5 = x_9 = \frac{1}{\sqrt{3}}$, all other $x_i = 0$ will also yield the same result. Since $\|S\| = 1$, we have $w(ST) > w(T)\|S\|$, as observed in Section 2.5.

The actual numerical range $W(TS)$ may be seen in Fig. 6 of Section 5.6.

We have already seen two Gersgorin-type estimates for the numerical range $W(A)$ in Section 5.2. In addition, we can use the following estimate

122   5. Finite Dimensions

for the numerical radius $w(A)$. Let

$$(5.3\text{-}4) \qquad b_i = \left(\sum_{j \neq i} |a_{ij}|^2 \right)^{1/2} + \left(\sum_{j \neq i} |a_{ji}|^2 \right)^{1/2}.$$

Denote by A_i the ith projection of A obtained by deleting the ith row and the ith column of A. Generally, let c_i denote $w(A_i)$. An estimate for $w(A)$ (see Abdurakmanov (1988), Notes and References) is given in terms of these c_i by

$$(5.3\text{-}5) \qquad w(A) = \min_i \left\{ \frac{|a_{ii}| + d_i}{2} + \frac{[|a_{ii} - d_i|^2 + b_i^2]^{\frac{1}{2}}}{2} \right\},$$

where d_i is any arbitrary constant so that $c_i \leq d_i$.

Let us illustrate the estimate (5.3-5) by a companion matrix example. The *companion matrix* of a polynomial $\lambda^n + p_1 \lambda^{n-1} + \cdots + p_n$ is given by

$$(5.3\text{-}6) \qquad A = \begin{bmatrix} 0 & 1 & 0 & 0 & \cdots & 0 \\ 0 & 0 & 1 & 0 & \cdots & 0 \\ \vdots & & & & & \\ 0 & 0 & 0 & 0 & \cdots & 1 \\ -p_n & -p_{n-1} & & & \cdots & -p_1 \end{bmatrix} = [a_{ij}].$$

In particular, A_n is rather simple for the companion matrix (5.3-6). In this case, the $(n-1) \times (n-1)$ matrix $A_n = S_{n-1}$, the $(n-1)$-dimensional shift, and we know that $w(A_n) = \cos \frac{\pi}{n}$. We can simplify (5.3-5) further by choosing $d_i = c_i$. Since $|a_{nn}| = |p_1|$, we get an estimate

$$(5.3\text{-}7) \qquad w(A) \leq \frac{|p_1| + \cos \frac{\pi}{n}}{2} + \frac{[(|p_1| - \cos \frac{\pi}{n})^2 + b_n^2]^{\frac{1}{2}}}{2}.$$

Example. For the polynomial $\lambda^3 - 3\lambda^2 + 4\lambda - 2$, we have

$$A = \begin{bmatrix} 0 & 1 & 0 \\ 0 & 0 & 1 \\ 2 & -4 & 3 \end{bmatrix}; \quad |p_1| = 3, \; b_n = \frac{(20)^{1/2} + 1}{2},$$

from which (5.3-7) becomes

$$w(A) \leq \frac{3.5 + ((2.5)^2 + ((2^2 + 4)^{1/2} + 1)^2)^{1/2}}{2} \cong 4.85.$$

We know that $w(A) \cong 3.77$ (see Fig. 3 in Section 5.6).

Clearly, one needs to calculate the two other bounds in (5.3-5), and optimize in d_i, to get the best bound by this method. Carrying out the former, we note (e.g., by numerical calculation; see Section 5.6) that $c_2 = w(A_2) \cong 3.30$ and $c_1 = w(A_1) \cong 3.62$, and again taking $d_2 = c_2$ and $d_1 = c_1$, (5.3-5) yields the estimates $w_{22} = 4.697$ and $w_{11} = 4.161$.

Notes and References for Section 5.3

An excellent exposition for the calculation of the numerical range of 0–1 matrices is given in

M. Marcus and B. N. Shure (1979). "The Numerical Range of Certain 0, 1-Matrices," *Lin. and Multilin. Alg.* **7**, 111–120.

The article contains additional results and a method of plotting the boundary of the numerical range. An extension of this result, and the example S_9 given earlier, were taken from

K. R. Davidson and J. A. R. Holbrook (1988). "Numerical Radii of Zero-One Matrices," *Michigan Math. J.* **35**, 261–267.

The estimate (5.3-5) is given in

A. A. Abdurakmanov (1988). "The Geometry of the Hausdorff Domain in Localization Problems for the Spectrum of Arbitrary Matrices," *Math. USSR Sbornik.* **59**, 39–51.

See also

C. Johnson (1974). "Gershgorin Sets and the Field of Values," *J. Math. Anal. Appl.* **45**, 416–419.

for the earlier numerical radius estimate

$$(5.3\text{-}8) \qquad w(A) \leqq \frac{1}{2} \max_i \left\{ \sum_j |a_{ij}| + \sum_j |a_{ji}| \right\}.$$

Mappings of the numerical radius can be found in

C. R. Johnson, I. M. Spitkovsky and S. Gottlieb (1994). "Inequalities Involving the Numerical Radius," *Linear and Multilinear Algebra* **37**, 13–24.

There it is shown that if f and g are polynomials with real coefficients and A is a 2×2 matrix with $\text{tr } A$ and $\det A$ real, then

$$w(f(A)g(A)) \leq w(f(A))w(g(A)).$$

Further, the following unsolved cases for 2×2 matrices are pointed out:
1. $A \in R^{2 \times 2}$, f and g have complex coefficients;
2. $A \in C^{2 \times 2}$, $f(z) = z$ and $g(z) = z^2$;
3. $A \in R^{n \times n}$, $f(z) = z^k$, $g(z) = z^m$, $k \neq m$, $2 < n \leq k + m$.

5.4 Hadamard Product

For any two matrices A and B of the same dimension, the Hadamard product is defined as the entrywise product. Thus, if $A, B \in M_{mn}$, $A = [a_{ij}]$, $B = [b_{ij}]$, their *Hadamard product* is given by

$$(5.4\text{-}1) \qquad A \circ B = [a_{ij}b_{ij}] \in M_{mn}.$$

124   5. Finite Dimensions

Evidently, $A \circ B = B \circ A$, $A \circ (B \circ C) = (A \circ B) \circ C$, $A \circ (B+C) = A \circ B + A \circ B$ for any $A, B, C \in M_{mn}$ and $(\alpha A) \circ B = \alpha(A \circ B)$ for any scalar α. Notice that if $A \circ B = A$, then all the elements of B should be equal to 1.

In this section, we study the numerical range of the product $A \circ B$, where $A, B \in M_n$ and one of them, say A, is normal. Many of the properties of $A \circ B$ can be derived from those of their Kronecker product (5.4-2) defined later. When $A \in M_{mn}$ and $B \in M_{pq}$ are two matrices of any dimension, $A = [a_{ij}]$, $B = [b_{ij}]$, their *Kronecker product* is defined as the matrix given by the blocks

$$(5.4\text{-}2) \qquad A \otimes B = \begin{bmatrix} a_{11}B & a_{12}B & \cdots & a_{1n}B \\ & \cdots & & \\ a_{n1}B & a_{n2}B & \cdots & a_{mn}B \end{bmatrix} \in M_{mp,nq}.$$

If

$$A = \begin{bmatrix} 2 & 3 & 4 \\ a & b & c \end{bmatrix} \quad \text{and} \quad B = \begin{bmatrix} \alpha \\ \beta \end{bmatrix},$$

then, for example, we have

$$(5.4\text{-}3) \qquad A \otimes B = \begin{bmatrix} 2\alpha & 3\alpha & 4\alpha \\ 2\beta & 3\beta & 4\beta \\ a\alpha & b\alpha & c\alpha \\ a\beta & b\beta & c\beta \end{bmatrix}.$$

The associated and distributive laws can be easily verified. Further, $(A \otimes B)^\top = A^\top \otimes B^\top$, $(A \otimes B)^* = A^* \otimes B^*$, $(\alpha A) \otimes B = A \otimes (\alpha B) = \alpha(A \otimes B)$. Also, if $A \otimes B$ and $C \otimes D$ permit ordinary matrix multiplication, we can verify that $(A \otimes B)(C \otimes D) = AC \otimes BD$. If $A, B, C, D \in M_n$, $A^{-1} = C$, $B^{-1} = D$, we can see that $(A \otimes B)(C \otimes D) = AC \otimes BD = I \otimes I$. Notice, however, that $A \otimes B$ is not necessarily $B \otimes A$.

Let us now consider the case when A and B are square matrices, $A, B \in M_n$.

Theorem 5.4-1. Let $A, B \in M_n$. Then
 (a) $\sigma(A \otimes B) = \{\lambda_i \mu_j,\ i, j = 1, 2, \ldots, n\}$, including algebraic multiplicities.
 (b) If A and B are positive Hermitian, then so is $A \otimes B$.

Proof. (a) Let $Ax = \lambda x$, $By = \lambda y$, $x \neq 0$, $y \neq 0$. Then $(A \otimes B)(x \otimes y) = Ax \otimes By = \lambda\mu x \otimes y$. Using the Schur triangularization, let U^*AU, V^*BV be upper triangular where U and V are unitary and the eigenvalues of A and B form the respective main diagonals

$$(U \otimes V)^*(A \otimes B)(U \otimes V) = U^*AU \otimes V^*V,$$

where both the factors are upper triangular.
 (b) $A = A^*$ and $B = B^*$. Then $(A \otimes B)^* = A^* \otimes B^* = A \otimes B$. □

5.4 Hadamard Product

Corollary 5.4-2. If A and B are positive definite matrices in M_n, then so is $A \otimes B$.

Proof. $A \otimes B$ is Hermitian, and now use Theorem 5.4-1. □

Theorem 5.4-3. Let $A \in M_n$, $B \in M_k$ be two square matrices. Then
(a) $\text{co}\,[W(A)W(B)] \subset W(A \otimes B)$.
(b) If A is normal, then $W(A \otimes B) = \text{co}\,[W(A)W(B)]$.

Proof. (a) Let $\langle Ax, x \rangle \in W(A)$ and $\langle By, y \rangle \in W(B)$, $\|x\| = \|y\| = 1$, with the corresponding inner products. Then
$$\langle x \otimes y, x \otimes y \rangle = \langle x, x \rangle \langle y, y \rangle = 1,$$
$$\langle Ax, x \rangle \langle By, y \rangle = \langle (A \otimes B)(x \otimes y), x \otimes y \rangle \in W(A \otimes B).$$
Thus $W(A)W(B) \subset W(A \otimes B)$. Since $W(A \otimes B)$ is always convex, we have $\text{co}\,[W(A)W(B)] \subset W(A \otimes B)$.

(b) Let A be normal and $U \in M_n$ a unitary matrix such that U^*AU is the diagonal matrix
$$D = \begin{bmatrix} \alpha_1 & 0 & 0 \\ 0 & \alpha_2 & \vdots \\ 0 & \cdots & \alpha_n \end{bmatrix}.$$
Since the numerical range is unitarily invariant, we have

(5.4-4)
$$W(A \otimes B) = W[(U \otimes I)^*(A \otimes B)(U \otimes I)]$$
$$= W(U^*AU \otimes B) = W(D \otimes B).$$

To calculate $W(D \otimes B)$, let us choose a general vector $x \otimes y$, where
$$x = \begin{bmatrix} x_1 \\ \vdots \\ x_n \end{bmatrix}, \quad y = \begin{bmatrix} y_1 \\ \vdots \\ y_k \end{bmatrix}, \quad \text{and} \quad x \otimes y = \begin{bmatrix} x_1 y \\ x_n y \end{bmatrix} \in \mathbb{C}^{nk}$$
and let us form the block matrix
$$D \otimes B = \begin{bmatrix} \alpha_1 & 0 & \cdots & 0 \\ 0 & \alpha_2 & & 0 \\ & & \vdots & \\ 0 & 0 & & \alpha_n \end{bmatrix} \quad B = \begin{bmatrix} \alpha_1 B & & \\ & \alpha_2 B & \\ & & \alpha_n B \end{bmatrix}.$$
Then

(5.4-5)
$$\langle (D \otimes B)(x \otimes y), x \otimes y \rangle = \sum_{i=1}^n \alpha_i x_i \bar{x}_i \langle By, y \rangle.$$

Let $J = \{i \in (1, 2, \ldots, n), x_i \neq 0\}$. Then
$$\langle (D \otimes B)(x \otimes y), x \otimes y \rangle = \sum_{i \in J} x_i \bar{x}_i \alpha_i \langle By, y \rangle = z.$$

Since $\alpha_i \langle By, y \rangle \in W(D)W(B)$ and $\sum_{i \in J} |x_i|^2 = 1$, we have z as a convex combination of elements of $W(D)W(B)$, and hence z belongs to co $[W(D) \cdot W(B)]$. The result (b) follows since $W(D) = W(A)$. □

We can now derive some results for the numerical range of the Hadamard product (5.4-1) of $A \circ B$ by regarding it as a submatrix of $A \otimes B$, as illustrated by the following example.

Example. Let

$$A = \begin{bmatrix} a_{11} & a_{12} \\ a_{21} & a_{22} \end{bmatrix}, \quad B = \begin{bmatrix} b_{11} & b_{12} \\ b_{21} & b_{22} \end{bmatrix}.$$

Then

(5.4-6) $$A \otimes B = \begin{bmatrix} \underline{a_{11}b_{11}} & a_{11}b_{12} & a_{12}b_{11} & a_{12}b_{12} \\ a_{11}b_{21} & \underline{a_{11}b_{22}} & a_{12}b_{21} & a_{12}b_{22} \\ a_{21}b_{11} & a_{21}b_{12} & \underline{a_{22}b_{11}} & a_{22}b_{12} \\ a_{21}b_{21} & a_{21}b_{22} & a_{22}b_{21} & \underline{a_{22}b_{22}} \end{bmatrix}.$$

Observe that the underlined elements form $A \circ B$. In general, for $A, B \in M_n$, the elements of $A \circ B$ appear in the columns and rows numbered $1, n+2, 2n+3, 3n+4, \ldots, n^2$ of $A \otimes B$. The following properties of $A \circ B$ can be deduced immediately from this inclusion.

Theorem 5.4-4. Let A and B be two commuting $n \times n$ matrices. Then
a) If A and B are Hermitian, so is $A \circ B$.
b) If A and B are positive semidefinite, so is $A \circ B$.
c) If A is normal, then $w(A \circ B) \le w(A)w(B)$.

Proof. $A \circ B$ is a submatrix of $A \otimes B$ and hence by the submatrix inclusion (Theorem 5.1-2), $W(A \circ C) \subset W(A \otimes B)$. For (a) observe that $W(A \otimes B)$ is real; for (b) $W(A \otimes B)$ is contained in the right half-plane; and for (c) observe that $W(A \circ B) \subset \mathrm{co}\,[W(A)W(B)]$. □

Notes and References for Section 5.4

The Hadamard product was initially studied by
J. Schur (1911). "Bemerkungen zur Theorie der Beschränkten Bilinearformen mit Unendlich vielen Veränderlichen," *J. Reine Angew. Math.* **140**, 1–28.
An excellent account and also an in-depth study of the Hadamard product and its applications can be found in
R. A. Horn (1990). "The Hadamard Product," *Proc. Symposia in Appl. Math.* **40**, 87–120.
Important results in the theory of Hadamard products can be seen in
T. Ando, R. A. Horn and C. R. Johnson (1987). "The Singular Values of a Hadamard Product: A Basic Inequality," *Lin. Multilin. Alg.* **21**, 345–365.

See also [HJ2].

5.5 Generalized Ranges

The numerical range found many generalizations with different applications in view. We will treat one of these generalizations, which has been extensively studied recently, mention some more, and then refer the reader to further studies in the literature.

C-Numerical Range

One of the generalized ranges that attracted a lot of attention is the C-numerical range. For any $A \in M_n$,

(5.5-1) $\qquad W_C(A) = \{\operatorname{tr}(CU^*AU),\ U \in U_n\},$

where U_n is the group of unitary matrices in M_n. Evidently, $W_C(A)$ is a unitary invariant and also $W_C(A) = W_A(C)$. Further, $W_C(A) = W_B(A)$ iff B is unitarily similar to C. We will first see when $W_C(A)$ is nontrivial, then study its numerical radius, and finally look at its convexity properties. Since $W_C(A)$ is trivial when C is a scalar, let us first look at some situations in which C is a scalar operator.

Theorem 5.5-1. *If C leaves invariant all m-dimensional subspaces of $\mathbb{C}^n$, $1 \leq m < n$, then C is a scalar.*

Proof. Let $\{e_1, e_2, \ldots, e_n\}$ be an orthonormal basis in $\mathbb{C}^n$. Then $Ce_i = \lambda_i e_i$, $i = 1, 2, \ldots, n$, as C leaves invariant all one-dimensional subspaces. Further $C(\sum_1^n e_i) = \mu(\sum_1^n e_i)$ and hence $\lambda_i = \mu$, $i = 1, 2, \ldots, n$. □

Theorem 5.5-2. *If C commutes with U^*AU for all $U \in U_n$, then either A or C is a scalar.*

Proof. Suppose that A is not a scalar and λ an eigenvalue of A with corresponding eigenspace M_λ, $\dim M_\lambda < n$. For any $U \in U^n$, U^*AU also has λ as an eigenvalue and the corresponding eigenspace is U^*M_λ. So for $y \in U^*M_\lambda$, $U^*AU(Cy) = C(U^*AU)y = CU^*AUU^*x$, (for some $x \in M_\lambda$) $= CU^*Ax = C\lambda U^*x = C\lambda y$. So $Cy \in U^*M_\lambda$. Thus C leaves U^*M_λ invariant. Since U is arbitrary, C leaves all the m-dimensional subspaces invariant. Hence, by Theorem 5.5-1, C is a scalar. □

Another way in which C (or A) can be a scalar is that $W_C(A)$ can be a constant.

Theorem 5.5-3. *If $\operatorname{tr}(CU^*AU) = $ constant for all $U \in U_n$, then C commutes with U^*AU, and hence either C or A is a scalar.*

Proof. For any $x \in R$, e^{xS} is unitary for any skew-symmetric matrix S. Consider the constant equal to $\text{tr}\,[Cu^{-xS}AU^{xS}]$. The derivative with respect to x at $x = 0$ is

$$0 = \text{tr}\,[(CU^*AU - U^*AUC)S].$$

Every matrix $B \in M_n$ can be written as a linear combination of two skew-symmetric matrices as

$$B = \frac{1}{2}(B - B^*) - \frac{i}{2}\left[\frac{i}{2}(B + B^*)\right].$$

Hence $\text{tr}\,[(CU^*AU - U^*AUC)B] = 0$ for all $B \in M_n$. So C commutes with U^*AU for all $U \in \mathcal{U}_n$ and hence is a scalar (or A is a scalar). □

The C-numerical radius $r_C(A)$ is given by

(5.5-2) $\qquad r_C(A) = \max\{|z| : z \in W_C(A)\}.$

Comparing it with the numerical radius $w(A)$, which is an equivalent norm, we first study the norm properties of $r_C(A)$. To this end, let us recall the following definitions.

A function $N : M_n \to R$ is called a *seminorm* if for all $A, B \in M_n$ and all $\alpha \in \mathbb{C}$,

(5.5-3)
$$N(A) \geq 0,$$
$$N(\alpha A) = |\alpha|N(A),$$
$$N(A + B) \leq N(A) + N(B).$$

A seminorm is a generalized matrix norm if it is positive definite or, equivalently,

$$N(A) > 0 \quad \text{whenever} \quad A \neq 0.$$

A generalized matrix norm is a matrix norm if, for all $A, B \in M_n$,

$$N(AB) \leq N(A)N(B).$$

Recalling the properties of $W(A)$, the best that we can expect is that $r_C(A)$ be a generalized matrix norm.

Theorem 5.5-4. *The C-numerical radius $r_C(A)$ is a generalized matrix norm if and only if C is not a scalar and has a nonzero trace.*

Proof. Let $r_C(A) = 0$. Then $\text{tr}\,(CU^*AU) = 0$ (a constant) for all $U \in \mathcal{U}_n$. Hence, by Theorem 5.5-3, either A or C is a scalar. Suppose that C is not a scalar. Then $A = \mu I$ and $r_C(A) = |\mu \,\text{tr}\,C| = 0$. If $\text{tr}\,C \neq 0$, we have $\mu = 0$ and hence $A = 0$. On the other hand, C is a scalar λI implies that for any $A \neq 0$, with $\text{tr}\,A = 0$, $r_C(A)) = |\lambda \,\text{tr}\,A| = 0$. Further, if $\text{tr}\,C = 0$, then $r_C(I) = 0$. □

Before we look at the convexity of $W_C(A)$, let us observe that $W_C(A)$ is rather intractable for arbitrary matrices C and A. Most of the known

5.5 Generalized Ranges

results correspond to C normal or Hermitian. When C is normal, by the unitary invariance of $W_C(A)$, we can assume $C = \text{diag}\{c_1, c_2, \ldots, c_n\}$, where $c_i \in \mathbb{C}$. Denoting the vector $(c_1, c_2, \ldots, c_n)$ by c, $W_C(A)$ is then usually written as $W_c(A)$. An equivalent formulation for $W_c(A)$ is the following,

$$(5.5\text{-}4) \qquad W_c(A) = \left\{ \sum_1^n c_i \langle Ae_i, e_i \rangle, \{e_i\}_{i=1}^n \in \Lambda_n \right\},$$

where Λ_n is the set of orthonormal bases for $\mathbb{C}^n$. In fact, this is the original formulation of $W_c(A)$. Let us first note that even when C and A are normal, $W_c(A)$ need not be convex.

Example. In $\mathbb{C}^3$, let $C = A = \text{diag}\{0, 1, i\}$. Then $c = (0, 1, i)$, and let us first show that for any o.n. basis $\{e_1, e_2, e_3\}$, we have

$$\sum_1^3 C_i \langle Ae_i, e_i \rangle = \langle Ae_2, e_2 \rangle + i \langle Ae_3, e_3 \rangle.$$

Let us choose

$$e_1 = \begin{bmatrix} 0 \\ 0 \\ 1 \end{bmatrix}, \quad e_2 = \begin{bmatrix} 0 \\ 1 \\ 0 \end{bmatrix}, \quad e_3 = \begin{bmatrix} 1 \\ 0 \\ 0 \end{bmatrix}$$

Then

$$Ae_2 = \begin{bmatrix} 0 \\ 1 \\ 0 \end{bmatrix}$$

and $\langle Ae_2, e_2 \rangle = 1$, $Ae_3 = 0 = \langle Ae_3, e_3 \rangle$. Thus $1 \in W_c(A)$. Let us now choose

$$e_1 = \begin{bmatrix} 1 \\ 0 \\ 0 \end{bmatrix}, \quad e_2 = \begin{bmatrix} 0 \\ e^{\frac{i\pi}{4}} \\ 0 \end{bmatrix}, \quad e_3 = \begin{bmatrix} 0 \\ 0 \\ 1 \end{bmatrix},$$

$Ae_2 = e_2$ and $\langle Ae_2, e_2 \rangle = i = \langle Ae_3, e_3 \rangle$. Then $2i \in W_c(A)$. However, $\frac{1}{2} + i \notin W_c(A)$ as shown now. For any

$$e_2 = \begin{bmatrix} x \\ y \\ z \end{bmatrix} \quad \text{and} \quad e_3 = \begin{bmatrix} x' \\ y' \\ z' \end{bmatrix},$$

let $|y|^2 + |z|^2 = \alpha^2$ and $|y'|^2 + |z'|^2 = \beta^2$. Then, we have

$$\sum_{i=1}^3 C_i \langle Ae_i, Ae_i \rangle \in W(\alpha^2 B + i\beta^2 B),$$

where B is the two-dimensional diagonal matrix

$$\begin{bmatrix} 1 & 0 \\ 0 & i \end{bmatrix}.$$

$W(\alpha^2 B)$ is the segment joining α^2 and $i\alpha^2$, and $W(i\beta^2 B)$ is the segment joining $i\beta^2$ and $-\beta^2$. It can be easily verified that $\frac{1}{2} + i \notin \alpha^2 W(B) + i\beta^2 W(B)$, as follows. If it were the case, then for any $x_\alpha + iy_\alpha \in \alpha^2 W(B)$ and $x_\beta + iy_\beta \in i\beta^2 W(B)$ we would have $x_\alpha + y_\alpha = \alpha^2$, $y_\alpha - x_\alpha = \beta^2$, $x_\alpha + x_\beta = \frac{1}{2}$, and $x_\beta + y_\beta = 1$, an incompatible system.

We will now study the convexity of $W_c(A)$ in some special cases.

Theorem 5.5-5. *If $A \in M_2$ and $\alpha = (\alpha_1, \alpha_2) \in \mathbb{C}^2$, then $W_\alpha(A)$ is convex.*

Proof. Let $z = \alpha_1 \langle Ax_1, x_1 \rangle + \alpha_2 \langle Ax_2, x_2 \rangle$ for any orthonormal basis $\{x_1, x_2\}$ of $\mathbb{C}^2$. Then

$$(5.5\text{-}5) \quad z = (\alpha_1 - \alpha_2) \left\langle \left[A - \left(\frac{1}{2} \operatorname{Re} A \right) I \right] x_1, x_1 \right\rangle + \frac{1}{2} (\alpha_1 + \alpha_2) \operatorname{tr} A,$$

which is an element of the numerical range of the matrix $(\alpha_1 - \alpha_2)A + (\alpha_2 \operatorname{tr} A)I$. □

We will now compare the numerical ranges corresponding to two vectors $\alpha = (\alpha_1, \alpha_2)$ and $\beta = (\beta_1, \beta_2)$ such that $\alpha_1 = a\beta_1 + (1-a)\beta_2$, $\alpha_2 = (1-a)\beta_1 + a\beta_2$, $0 \le a \le 1$. We say that (α_1, α_2) is obtained from (β_1, β_2) by *pinching*.

Theorem 5.5-6. *If (α_1, α_2) is obtained by pinching (β_1, β_2), then*

$$(5.5\text{-}6) \quad W_\alpha(A) \subset W_\beta(A).$$

Proof. As in Theorem 5.5-5, let us write

$$W_{(\alpha_1, \alpha_2)}(A) = (\alpha_1 - \alpha_2)W(B) + \frac{1}{2}(\alpha_1 + \alpha_2)(\operatorname{tr} A)$$

$$= (2a - 1)(\beta_1 - \beta_2)W(B) + \frac{1}{2}(\beta_1 + \beta_2)(\operatorname{tr}(A))$$

and

$$W_{(\beta_1, \beta_2)}(A) = (\beta_1 - \beta_2)W(B) + \frac{1}{2}(\beta_1 + \beta_2)(\operatorname{tr} A),$$

where $B = A - \frac{1}{2}(\operatorname{tr} A)I$. By the elliptic range theorem, $(\beta_1 - \beta_2)W(B)$ is an ellipse and $\operatorname{tr} B = 0$. Hence $(\beta_1 - \beta_2)W(B)$ is symmetric about the origin. Since $0 \le a \le 1$, we have $-1 \le (2a - 1) \le 1$, and hence

$$(2a - 1)W(B) \subset W(B). \quad \square$$

Theorem 5.5-6 can be generalized to two vectors $\alpha, \beta \in \mathbb{C}^n$, when α is obtained from β by pinching, that is, when two components α_i, α_j are replaced by $a\beta_i + (1-a)\beta_j$ and $(1-a)\beta_i + a\beta_j$, where $0 \le a \le 1$, all other components of α remaining unchanged.

Theorem 5.5-7. *If α is obtained from β by pinching, then $W_\alpha(A) \subset W_\beta(A)$. If α is obtained from β by a finite number of pinchings, then $W_\alpha(A) \subset W_\beta(A)$.*

Proof. Let V be the subspace spanned by α_i, α_j and P the projection on V. Then $W_\alpha(A)$ and $W_\beta(A)$ consist of

$$\sum_{\substack{k=1 \\ k \ne i,j}}^{n} \alpha_k \langle Ax_k, x_k \rangle + W_{(\alpha_i,\alpha_j)}(PA)$$

and

$$\sum_{\substack{k=1 \\ k \ne i,j}}^{n} \alpha_k \langle Ax_k, x_k \rangle + W_{(\beta_1,\beta_2)}(PA).$$

However, $W_{(\alpha_i,\alpha_j)}(PA) \subset W_{(\beta_i,\beta_j)}(PA)$ and hence

$$W_\alpha(A) \subset W_\beta(A).$$

For the second part of the theorem, apply the first part repeatedly. □

Let us now specialize to the case when α and β are real vectors. Let β be ordered $(\beta_1 \ge \beta_2 \ge \cdots \ge \beta_n)$ and $\sum_{i=1}^{n} \alpha_i = \sum_{i=1}^{n} \beta_i$. Then, we can easily see that

$$(5.5\text{-}7) \qquad \sum_{i=1}^{k} \alpha_i \le \sum_{i=1}^{k} \beta_i, \quad 1 \le k \le n.$$

From (5.5-7) we may obtain an interesting relation between pinching and doubly stochastic matrices when α and β are real vectors and β is ordered. Recall that a matrix $A \in M_n$ is called *doubly stochastic* if all of its entries are nonnegative and all row sums and column sums are 1.

Theorem 5.5-8. *Let α, β be real n-vectors, $\sum_1^n \alpha_i = \sum_1^n \beta_i$, with β ordered, $\beta_1 \ge \beta_2 \ge \cdots \ge \beta_n$. If α is obtained from β by pinching a finite number of times, then there exists a doubly stochastic matrix S such that $\alpha = S\beta$.*

Proof. Let us first see the effect of one pinching. If γ is obtained from β by pinching β_i, β_j, we have

$$\gamma_i = a\beta_i + (1-a)\beta_j,$$
$$\gamma_j = (1-a)\beta_i + a\beta_j.$$

If $i < j$, we have

(5.5-8) $$\gamma = \begin{bmatrix} 1 & 0 & & & & & \\ 0 & 1 & & & & & \\ & & \ddots & & & & \\ & & & a & 1-a & & \\ & & & 1-a & a & & \\ & & & & & \ddots & \end{bmatrix} \beta.$$

The matrix involved is doubly stochastic and hence $\alpha = S\beta$, where S is the product of doubly stochastic matrices and is doubly stochastic. □

Let us now use the above development to study the case of $W_\alpha(A)$ when α is a real vector $\alpha = (\alpha_1, \ldots, \alpha_n) \in \mathbb{R}^n$. Let us write for any $b, c \in \mathbb{R}^n$, $b \prec c$ if there exists a doubly stochastic matrix S such that $b = cS$. For example, in Theorem 5.5-8, we could write $\alpha \prec \beta$, i.e., α is majorized by β.

Theorem 5.5-9. *For $A, C \in M_n$ and C Hermitian, $W_C(A)$ is convex.*

Proof. Since $W_C(A)$ is a unitary invariant, we may take C as the diagonal matrix $[c]$, where $C \in R^n$. Let $\{U^*[c]U : U \in \mathcal{U}_n\}$ be denoted by $M(c)$. Then $W_C(A) = \{\text{tr}(Ax), x \in M(c)\} \subset \{\text{tr}(Ax), x \in \text{conv}\, M(c)\}$. We will now make use of a result (see Notes at the end of this section) which, with the majorization notation described above shows that $\text{conv}\, M(c) = \{U^*[b]U, U \in \mathcal{U}_n, b \prec c\}$. We will then have $\text{tr}(Ax) \in W_b(A) \subset W_c(A)$ by Theorem 5.5-7. Hence $W_c(A) = \{\text{tr}(Ax), x \in \text{conv}\, M(c)\}$, which is convex. □

Some other variations on the classical numerical range are the following.

k-Numerical Range

Let us note that the k-numerical range $W_k(A)$ defined as

$$W_k(A) = \left\{ z : z = \sum_{i=1}^{k} \langle Ax_i, x_i \rangle, \text{ where } \{x_1, \ldots, x_k\} \text{ are } k \text{ orthonormal vectors in } \mathbb{C}^n \right\}$$

is a special case of $W_c(A)$, where $c \in \mathbb{R}^n$.

F-Numerical Range

If $\|\ \|$ is any norm in C^n, the norm induced in its dual space is given by

$$\|y\|^* = \sup\{|y^*x|, x \in \mathbb{C}^n, \|x\| = 1\}.$$

The Bauer field of values, also known more generally as the *spatial numerical range* $V(A)$, is given by

$$F(A) = \{y^*Ax, x \in C^n, \|x\| = \|y^*\| = |y^*x| = 1\}.$$

This generalized numerical range was discussed in Section 1.6. In particular, $F(A)$ is not necessarily convex.

Algebraic Numerical Ranges

A generalization of $F(A)$ called the algebraic numerical range is obtained in the following manner. For any $A \in M_n$, let $\|A\| = \sup\{\|Ax\|, x \in \mathbb{C}^n, \|x\| = 1\}$. Let $(M_n, \|\ \|^*)$ be the dual space given by

$$\|B\|^* = \sup\{|\text{tr}\,(BA)| : A \in (M_n, \|\ \|), \|A\| = 1\}.$$

The algebraic numerical range is taken to be

$$\tau(A) = \{\text{tr}\,(BA) : \|B^*\| = \text{tr}\,B = 1\}.$$

It is known (see the References at the end of this section) that $\tau(A) = \bigcap D[\gamma, \rho]$, where $D[\gamma, \rho]$ is the disk $\{z : |z - \gamma| \leq \rho\}$, and the intersection is over all pairs γ, ρ for which $\|A - \gamma I\| \leq \rho$. In general, $\tau(A)$ is the convex hull of $F(A)$.

Example. Let the norm chosen be $\|\ \|_\infty$. Then, for any $A = [a_{ij}] \in M_n$, let $D_i(A)$ denote the Gersgorin disk (see Section 5.2) $D_i(A) = D[a_{ii}, \rho_{ii}]$, where $\rho_{ii} = \sum_{i \neq k} |a_{ik}|$. Then

$$\tau(A) = \text{conv}\left(\bigcup_1^n D_i\right).$$

For algebraic numerical ranges for the infinite-dimensional case, such as Banach algebras, see [BD]. Those are usually denoted as $v(A)$ in those contexts.

M-Numerical Range

A variation on algebraic numerical ranges motivated by applications in stability analysis is given by

$$\tau_M(A) = \bigcap D[\gamma, \rho],$$

where the intersection is over all pairs γ, ρ for which $\|(A - \gamma I)^k\| \leq M\rho^k$, where $M \geq 1$ is fixed and $k = 1, 2, \ldots$. Some elementary properties that can be easily verified are

$$\tau_M(A) \text{ is compact and convex,}$$
$$\tau_M(\alpha I + \beta A) = \alpha + \beta\tau_M(A),$$
$$\tau_M(A) \subset \tau_N(A) \text{ if } 1 \leq N \leq M,$$
$$\sigma(A) \subset \tau_M(A).$$

Usually, $M > 1$ in applications. The motivation for $\tau_M(A)$ is to obtain containment sets for $\sigma(A)$ tighter than those provided just by $W(A)$.

Restricted Numerical Ranges

The subset of the complex numbers consisting of

$$W_S(T) = \{\langle Tx, x\rangle, \|x\| = 1, x \in S\},$$

where $S \subset H$ is a prescribed set, is called a restricted numerical range. The δ-*numerical range* $W_\delta(T)$ is an example of this: $S = \{x : \|Tx\| \geq \delta\}$. When S is the whole unit sphere, we have the classical numerical range.

Let us now consider two properties on S that will guarantee that $W_S(T)$ will be convex:

(i) $x \in S$ implies $\alpha x \in S$ if $|\alpha| = 1$;
(ii) if $x, y \in S$, then for every $r > 0$, either

$$\frac{x+ry}{\|x+ry\|} \in S \quad \text{or} \quad \frac{x-ry}{\|x-ry\|} \in S.$$

Theorem 5.5-10. *If S satisfies the properties (i) and (ii), then $W_S(T)$ is convex.*

Proof. Let $x, y \in S$, $\|x\| = \|y\| = 1$. Then, for $0 < t < 1$, consider

(5.5-9) $$\frac{\langle T(x+\alpha y), x+\alpha y\rangle}{\|x+\alpha y\|^2} = t\langle Tx, x\rangle + (1-t)\langle Ty, y\rangle,$$

the condition for convexity of S. Simplifying (5.5-9), we get an expression of the form $|\alpha|^2 + a\alpha + b\bar{\alpha} - \frac{1-t}{t} = 0$ where a and b are complex numbers depending only on T, x, y but not on α. Separating the real and imaginary parts and observing that $\frac{1-t}{t} > 0$, we get the equation of a line passing through the origin. Thus there exist two values of α satisfying (5.5-9), namely the intersections of the line with the circle. The assumptions on S now guarantee that one of these belongs to S. □

Corollary 5.5-11. $W_\delta(T)$ *is convex.*

Proof. Notice that $\{x : \|Tx\| \geq \delta\} = \{x : \|\sqrt{T^*T}\| \geq \delta\}$ and $\sqrt{T^*T}$ is selfadjoint. In general, a set $S = \{x \in H, \|x\| = 1, \langle Tx, x\rangle \geq \delta\}$, in the case of T selfadjoint, satisfies conditions (i) and (ii). This is seen as follows. For

$r > 0$, $x, y \in S$, we have $\langle T(x \pm ry), x \pm ry \rangle = 1$, $\langle Tx, x \rangle + r^2 \langle Ty, y \rangle \pm 2r \operatorname{Re} \langle Tx, y \rangle$. Thus

$$\frac{\langle T(x \pm ry), x \pm ry \rangle}{\|x + ry\|^2} \geq \delta \pm \frac{2r \operatorname{Re} \langle Tx - \delta x, y \rangle}{\|x + ry\|^2}.$$

Depending on the sign of $\operatorname{Re} \langle Tx - \delta x, y \rangle$, we can say that

$$\frac{x + ry}{\|x + ry\|} \quad \text{or} \quad \frac{x - ry}{\|x - ry\|} \in S.$$

Corollary 5.5-11 follows by reduction to this case and Theorem 5.5-10. □

Symmetric Numerical Range

Let $\|\ \|$ be a norm on $\mathbb{C}^n$ and $\|\ \|^*$ be the dual norm induced. Let $S = \{x \in \mathbb{C}^n, \|x\| = 1\}$ and $S^* = \{y \in \mathbb{C}^n, \|y\|^* = 1\}$ denote the Cartesian product $S \times S^*$ by π.

The symmetric numerical range of a matrix A is defined as

$$Z(A) = \operatorname{conv} \left\{ \frac{1}{2}(y^*Ax + x^*Ay), (x, y) \in \pi \right\}.$$

$Z(A)$ satisfies all the usual properties (convex, spectral inclusion, $Z(A + B) \subset Z(A) + Z(B)$, $Z(\lambda A) = \lambda Z(A)$ of a numerical range (see the Notes at the end of this section).

Notes and References for Section 5.5

C-numerical ranges have been studied extensively. A recent review of the results and some open problems can be seen in
C. K. Li (1994). "C-Numerical Ranges and C-Numerical Radii," *Lin. Multilin. Alg.* **37**, 51–82.
Another review article is that of
M. Goldberg (1979). "On Certain Finite Dimensional Numerical Ranges and Numerical Radii," *Lin. Multilin. Alg.* **7**, 329–342.

Inequalities involving the C-numerical radius can be found in
M. Goldberg and E. G. Straus (1979). "Norm Properties of C-Numerical Radii," *Lin. Alg. Appl.* **24**, 113–131.

The geometry of $W_C(A)$ has been studied by various authors. A review can be found in
N. Bebiano and J. da Providencia (1994). "Some Geometrical Properties of the Numerical Range of a Normal Matrix," *Lin. Multilin. Alg.* **37**, 83–92.
The technique of pinching and its use in inclusion relations can be found in
M. Goldberg and E. G. Straus (1977). "Elementary Inclusion Relations for Generalized Numerical Ranges," *Lin. Alg. Appl.* **18**, 11–24.

The convexity of $W_c(A)$ and Theorem 5.5-9 when c is real or a rotation of a real vector was shown by

R. Westwick (1975). "A Theorem on Numerical Range," *Lin. Multilin. Alg.* **2**, 311–315.

The simplified version we gave is due to

Y. T. Poon (1980). "Another Proof of a Result of Westwick," *Lin. Multilin. Alg.* **9**, 35–37.

$W_C(A)$, although not convex for normal C, is star-shaped. This was shown in

N. K. Tsing (1981). "On the Shape of the Generalized Numerical Range," *Lin. Multilin. Alg.* **10**, 173–182.

An infinite-dimensional version of the above theorem appears in

M. S. Jones (1991), "A Note on the Shape of the Generalized C-Numerical Range," *Linear and Multilinear Algebra* **31**, 81–84.

An earlier version can be found in

G. Hughes (1990). "A Note on the Shape of the Generalized Numerical Range," *Linear and Multilinear Algebra*, **26**, 43–47.

It has been announced in

N. K. Tsing and W. S. Cheung (1996). "Star-Shapedness of the Generalized Numerical Ranges," in *Abstracts 3rd Workshop on Numerical Ranges and Numerical Radii* (T. Ando and K. Okubo, eds.), Sapporo, Japan.

that $W_C(A)$ is always star-shaped with center $(\operatorname{tr} A)(\operatorname{tr} C)/n$.

The M-numerical range and its applications can be seen in

M. N. Spijker (1993). "Numerical Ranges and Stability Estimates," *Appl Num. Math.* **13**, 241–249.

H. W. J. Lenferink and M. N. Spijker (1990). "A Generalization of the Numerical Range of a Matrix," *Lin. Alg. Appl.* **140**, 251–266.

The δ-numerical range was first studied by

J. Stampfli (1970). "The Norm of a Derivation," *Pacific Journal of Math.* **33**, 737–747.

The convexity of W_δ was proved by

J. Kyle (1977). "W_δ is Convex," *Pacific Journal of Math.* **72**, 483–485.

The generalization to the restricted numerical range was given in

K. Das, S. Mazumdar and B. Sims (1987). "Restricted Numerical Range and Weak Convergence on the Boundary of the Numerical Range," *J. Math. Phys. Sci.* **21**, 35–41.

The k-numerical range first mentioned by Halmos in [H] was studied and generalized in

Y. Poon (1980). "The Generalized k-Numerical Range," *Linear and Multilinear Algebra* **9**, 181–186.

M. Marcus (1979). "Some Combinatorial Aspects of the Generalized Numerical Range," *Ann. New York Acad. Sci.* **319**, 368–376.

Y. Au-Yeung and N. Tsing (1983). "A Conjecture of Marcus on the Generalized Numerical Range," *Linear and Multilinear Algebra* **14**, 235–239.

W. Man (1987). "The Convexity of the Generalized Numerical Range," *Linear and Multilinear Algebra* **20**, 229–245.

For details on the symmetric numerical range, see

B. D. Saunders and H. Schneider (1976). "A Symmetric Numerical Range for Matrices," *Numer. Math.* **26**, 99–105.

5.6 $W(A)$ Computation

As an experimental tool, it is convenient to have a code to produce $W(A)$ graphics. There are essentially two ways to go about this: either from without or from within.

Approximation from without uses the elementary idea that the boundary $\partial W(A)$ may be traced by computing the maximum eigenvalue $\lambda_\theta^{\max}$ and the associated eigenvector x_θ of the real part of $e^{i\theta}A$ as θ runs over a reasonably finite discretization of $0 \leq \theta \leq 2\pi$. The point $\langle x_\theta, \text{Re}\,(Ae^{i\theta})x_\theta\rangle/\langle x_\theta, x_\theta\rangle$ will be a boundary point of $W(A)$. The graph of $\partial W(A)$ can also be constructed using only the values $\lambda_\theta^{\max}$. In that case, we have

$$(5.6\text{-}1) \qquad W(A) \subset \bigcap_{0 \leqq \theta \leqq \pi} [\text{half-plane} : e^{-i\theta}\{z : \text{Re}\,z \leqq \lambda_\theta^{\max}\}].$$

The computation thus reduces to a subroutine for $\lambda_{\max}$.

Approximation from within uses the fact (Section 1.1) that $W(A)$ is the union of numerical ranges of two-dimensional submatrices. More precisely, if A is an $n \times n$ general complex matrix, then $W(A)$ is the union of all $W(A_{uv})$, where

$$A_{uv} = \begin{bmatrix} \langle Au, u\rangle & \langle Av, u\rangle \\ \langle Au, v\rangle & \langle Av, v\rangle \end{bmatrix},$$

where u and v run over all pairs of real orthonormal vectors. The computation thus reduces to generating a reasonably complete set of random orthonormal pairs u, v and plotting the elliptical disks $W(A_{uv})$ on top of each other, thus shading the interior of $W(A)$.

In the figures that follow, we preferred the outside-in approach, which takes efficient computational advantage of the bounded convexity of $W(A)$. On the other hand, we would like to mention that the inside-out approach is theoretically more interesting because it constructs all two-dimensional real compressions of A.

For finite matrices, the outside-in approach has some clear advantages of coding ease and efficiencies. On the other hand, we would like to note that if one wanted to extend these $W(A)$ computations to unbounded operators A, the inside-out approach would still be valid, whereas the outside-in approach would have to be modified because of potential discrepancies between the domains $\mathcal{D}(A)$ and $\mathcal{D}(A^*)$ when forming Re A.

138   5. Finite Dimensions

Figures 1, 2, and 3 are the numerical ranges $W(A)$ of the three examples 1, 2, and 3 of Section 5.2, respectively. Fig. 4 is $W(A)$ for the 0–1 example A of Section 5.3. Fig. 5 is $W(T)$ for $T = S^3 + S^7$ of Section 5.3; $W(S)$ is exactly the same. Fig. 6 is $W(TS)$ of that section. Because the numerical ranges of A and its adjoint A^* are the same for real matrices, Fig. 5 and 6 are also those for the T and TS of Section 2.5.

Figure 7 is $W(A)$ for a simple, nonsymmetric, banded 0–1 matrix which might come from a discretization of a first-order differential equation or from some Toeplitz matrix application. There the matrix A is

$$A = \begin{bmatrix} 0 & 1 & 0 & 0 & 0 \\ 0 & 0 & 1 & 0 & 0 \\ 1 & 0 & 0 & 1 & 0 \\ 0 & 1 & 0 & 0 & 1 \\ 0 & 0 & 1 & 0 & 0 \end{bmatrix}.$$

Compare with Fig. 8, in which the band structure is missing, where A is

$$A = \begin{bmatrix} 0 & 0 & 1 & 0 & 0 \\ 1 & 0 & 0 & 0 & 0 \\ 0 & 0 & 0 & 0 & 1 \\ 0 & 1 & 0 & 0 & 0 \\ 0 & 0 & 0 & 1 & 0 \end{bmatrix}.$$

Notice $w(A) = 1$ for the latter, in accordance with Section 5.3. Figure 9 is the numerical range $W(A)$ of

$$A = \begin{bmatrix} 4 & 0 & 0 & -1 \\ -1 & 4 & 0 & 0 \\ 0 & -1 & 4 & 0 \\ 0 & 0 & -1 & 4 \end{bmatrix}.$$

which might come from the discretization of a second-order partial differential equation.

5.6 $W(A)$ Computation   139

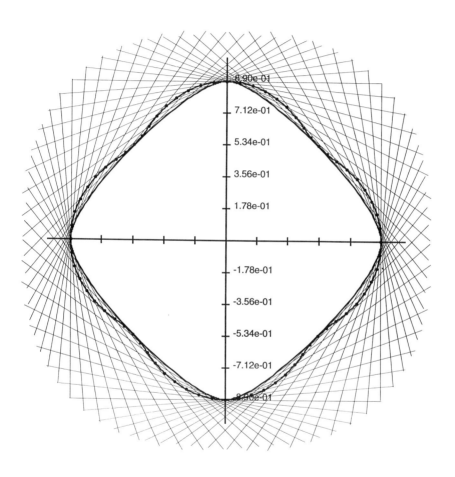

Figure 1. $W(A)$ for $A = \begin{bmatrix} 0 & 1 & 0 & 0 \\ 0 & 0 & 1 & 0 \\ 0 & 0 & 0 & 1 \\ 1/2 & 0 & 0 & 0 \end{bmatrix}$

140   5. Finite Dimensions

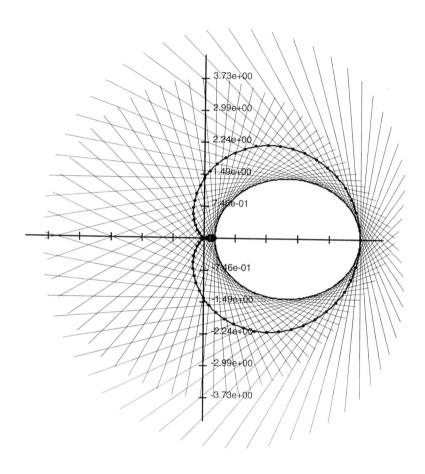

Figure 2. $W(A)$ for $A = \begin{bmatrix} 1 & 2 & 0 \\ 0 & 2 & 2 \\ 0 & 0 & 3 \end{bmatrix}$

5.6 $W(A)$ Computation 141

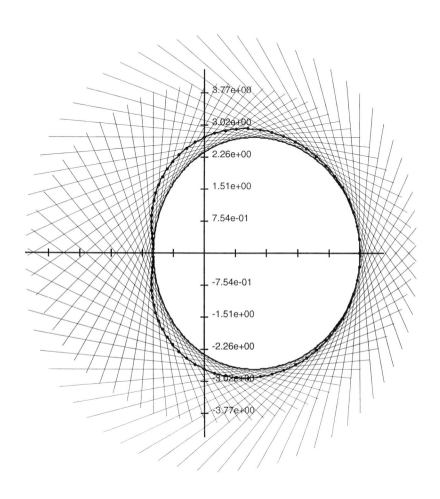

Figure 3. $W(A)$ for $A = \begin{bmatrix} 0 & 1 & 0 \\ 0 & 0 & 1 \\ 2 & -4 & 3 \end{bmatrix}$

142   5. Finite Dimensions

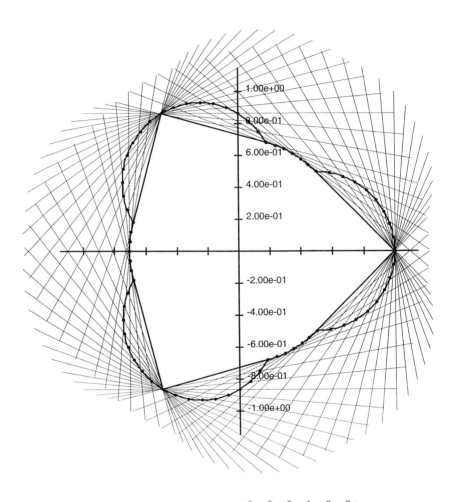

Figure 4. $W(A)$ for $A = \begin{bmatrix} 0 & 0 & 0 & 1 & 0 & 0 \\ 0 & 0 & 0 & 0 & 0 & 0 \\ 1 & 0 & 0 & 0 & 0 & 0 \\ 0 & 0 & 1 & 0 & 0 & 0 \\ 0 & 1 & 0 & 0 & 0 & 0 \\ 0 & 0 & 0 & 0 & 1 & 0 \end{bmatrix}$

5.6 $W(A)$ Computation 143

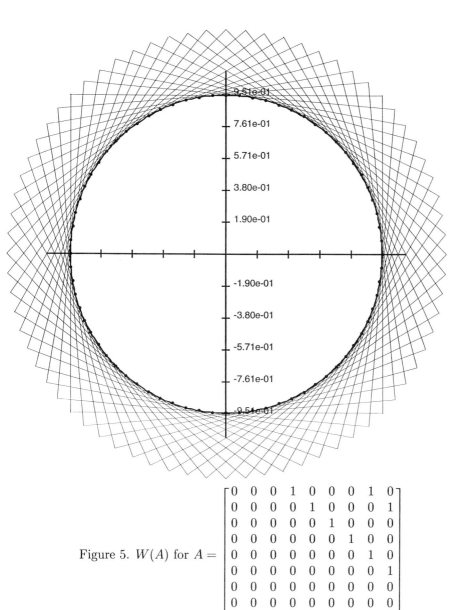

Figure 5. $W(A)$ for $A = \begin{bmatrix} 0 & 0 & 0 & 1 & 0 & 0 & 0 & 1 & 0 \\ 0 & 0 & 0 & 0 & 1 & 0 & 0 & 0 & 1 \\ 0 & 0 & 0 & 0 & 0 & 1 & 0 & 0 & 0 \\ 0 & 0 & 0 & 0 & 0 & 0 & 1 & 0 & 0 \\ 0 & 0 & 0 & 0 & 0 & 0 & 0 & 1 & 0 \\ 0 & 0 & 0 & 0 & 0 & 0 & 0 & 0 & 1 \\ 0 & 0 & 0 & 0 & 0 & 0 & 0 & 0 & 0 \\ 0 & 0 & 0 & 0 & 0 & 0 & 0 & 0 & 0 \\ 0 & 0 & 0 & 0 & 0 & 0 & 0 & 0 & 0 \end{bmatrix}$

144   5. Finite Dimensions

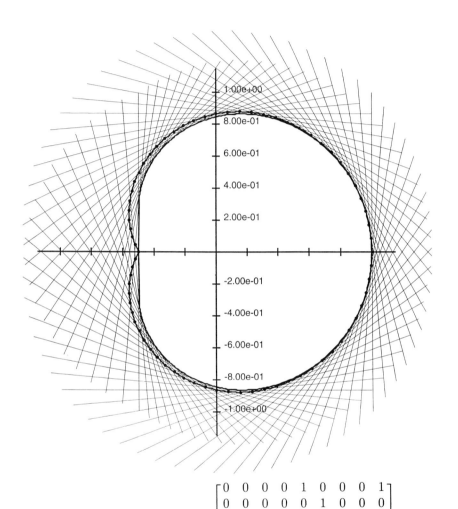

Figure 6. $W(A)$ for $A = \begin{bmatrix} 0 & 0 & 0 & 0 & 1 & 0 & 0 & 0 & 1 \\ 0 & 0 & 0 & 0 & 0 & 1 & 0 & 0 & 0 \\ 0 & 0 & 0 & 0 & 0 & 0 & 1 & 0 & 0 \\ 0 & 0 & 0 & 0 & 0 & 0 & 0 & 1 & 0 \\ 0 & 0 & 0 & 0 & 0 & 0 & 0 & 0 & 1 \\ 0 & 0 & 0 & 0 & 0 & 0 & 0 & 0 & 0 \\ 0 & 0 & 0 & 0 & 0 & 0 & 0 & 0 & 0 \\ 0 & 0 & 0 & 0 & 0 & 0 & 0 & 0 & 0 \\ 0 & 0 & 0 & 0 & 0 & 0 & 0 & 0 & 0 \end{bmatrix}$

5.6 $W(A)$ Computation   145

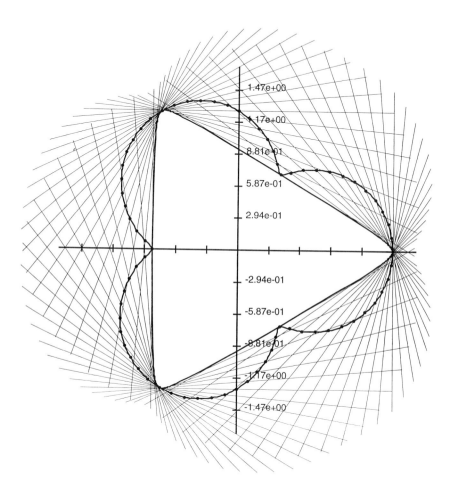

Figure 7. $W(A)$ for $A = \begin{bmatrix} 0 & 1 & 0 & 0 & 0 \\ 0 & 0 & 1 & 0 & 0 \\ 1 & 0 & 0 & 1 & 0 \\ 0 & 1 & 0 & 0 & 1 \\ 0 & 0 & 1 & 0 & 0 \end{bmatrix}$

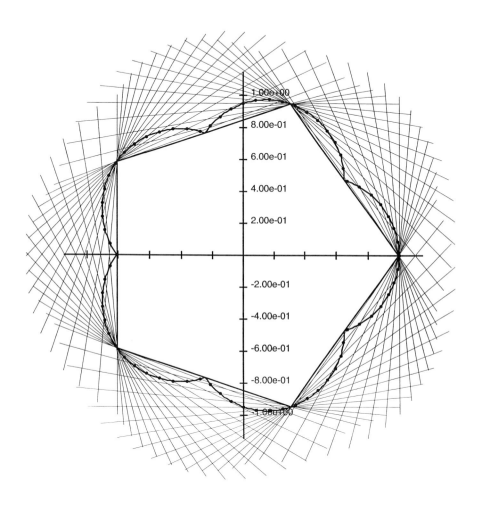

Figure 8. $W(A)$ for $A = \begin{bmatrix} 0 & 0 & 1 & 0 & 0 \\ 1 & 0 & 0 & 0 & 0 \\ 0 & 0 & 0 & 0 & 1 \\ 0 & 1 & 0 & 0 & 0 \\ 0 & 0 & 0 & 1 & 0 \end{bmatrix}$

5.6 $W(A)$ Computation   147

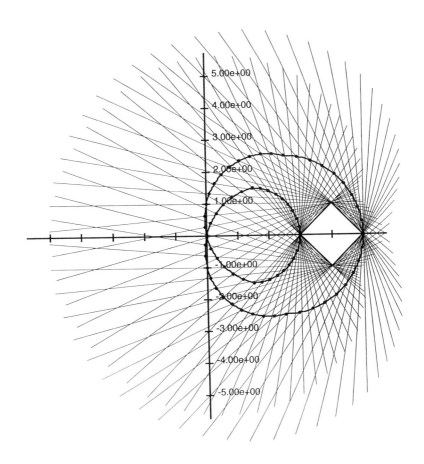

Figure 9. $W(A)$ for $A = \begin{bmatrix} 4 & 0 & 0 & -1 \\ -1 & 4 & 0 & 0 \\ 0 & -1 & 4 & 0 \\ 0 & 0 & -1 & 4 \end{bmatrix}$

Notes and References for Section 5.6

The $W(A)$ computation from outside-in is due to
C. Johnson (1978). "Numerical Determination of the Field of Values of a General Complex Matrix," *SIAM J. Numer. Anal.* **15**, 595–602.

The $W(A)$ computation from inside-out is due to
M. Marcus (1987). "Computer Generated Numerical Ranges and Some Resulting Theorems," *Linear and Multilin. Alg.* **20**, 121–157.

Also see [M] for more information on $W(A)$ and other matrix computations on personal computers.

The code and graphics for the figures produced here were written by Dr. G. Sartoris, using a variant of the outside-in supporting-hyperplane approach. One can shade the outside with tangent lines if so desired. We chose to do so to better highlight $W(A)$.

$W(A)$ will be x-axis symmetric for all real matrices A, as in the figures presented here. Minor modifications enable similar computation of $W(A)$ for complex matrices A.

Endnotes for Chapter 5

The core of the numerical range theory lies in the Hilbert space setting. Basic finite-dimensional theory is included in the Hilbert space theory, which, historically speaking, prompted generalizations to the Banach space situations.

Numerical range theory in finite dimensions, beyond its intrinsic interest as a part of matrix analysis, mainly has applications as a goal. As we have presented it, Section 5.1 gives applications in matrix theory, and Sections 5.2 and 5.3 are devoted to methods for finding bounds on the spectrum and on the numerical range. The theme of Section 5.4, the Hadamard product, originated in applications to differential equations. Many of the generalizations of numerical ranges in Section 5.5 were motivated by applications, and we presume and hope that some more generalizations will appear in the future with specific applications in view.

We may synthesize a tentative suggestion: an important emphasis in numerical range theory for finite dimensions in the immediate future should be to connect more specifically with the enormous number of new and important matrix classes now coming out of applications and matrix computations. For example, each basic matrix iterative method (e.g., Jacobi, Gauss–Seidel, successive over relaxation), to say nothing of all the important recent variations on the conjugate gradient method (such as GMRES, Orthomin, Lanzcos methods), all implicit or semi-implicit methods (such as ADI, etc.,...) coming out of discretizations of two- and three-dimensional physical problems in computational fluid dynamics, computational physics, computational chemistry, computational engineering in general, all classes of sparse and other matrices coming out of econometrics and statistics, should become candidates for connection to the theory of the numerical range. This would increase the richness of the subject of numerical range and aid understanding of the properties of each specific matrix class now being used extensively in applications and computations.

For example, to illustrate this synthesis, we chose the matrix A of Fig. 7 to be representative of the type studied by

L. N. Trefethen (1990). "Approximation Theory and Numerical Linear Algebra," in *Algorithms for Approximation*. II, eds. J. Mason and M. Cox, Chapman, London, 336–360.

in the investigation of pseudo-eigenvalues (see Section 4.6 here) and of the type studied by

M. Eiermann (1993). "Fields of Values and Iterative Methods," *Linear Alg. & Applic.* **180**, 167–197.

in the investigation of matrix classes arising from linear solver iterative schemes (see Fig. 2b there). Figs. 8 and 9 are numerical ranges for matrices A we took from

D. M. Young (1971). *Iterative Solution of large Linear Systems*, Academic Press, New York.

Specifically, Fig. 8 is $W(A)$ for A on p. 426, a matrix that is irreducible $CO(2,3)$ but is not a $CO^*(2,3)$ matrix, in the terminology of property-A matrices discussed there. The $W(A)$ of Fig. 9 is that of matrix A of Exercise 11, p. 431, a matrix that may be permuted so that $P^{-1}AP$ is a $CO(2,2)$ matrix.

We may mention another example, the 113×113 sparseness matrix A of K. Gustafson and R. Hartman (1985). "Graph Theory and Fluid Dynamics," *SIAM J. Alg. Disc. Math.* **6**, 643–656,

see also

K. Gustafson (1987). *Partial Differential Equations*, 2nd Ed., Wiley, New York, 311.

which would certainly have an interesting numerical range. That matrix represents the support, found by graph theoretic methods, of a basis for discrete solenoidal vectors for a finite-element scheme for the weak solution of a rather simple partial differential equation coming from linearized fluid dynamics.

These examples carry no special significance in themselves and were quickly chosen just to illustrate the synthesis that we are advancing in this endnote. During the twenty-five years since Young's book (1971), many more interesting classes of matrices have been investigated for iterative solutions. Similarly, for partial differential equations, finite-element methods have augmented finite-difference methods in applications, yielding many interesting classes of large sparse matrices. Depending on the grid generation method, some of these have good pattern structure, and some do not.

6

Operator Classes

Introduction

The special properties of the numerical range of normal operators led to the creation of new operator classes. In particular, three classes of operators attracted a lot of attention. Each of these inherits one or more of the numerical range properties of the normal operator.

An operator is *normaloid* if its norm and numerical radius are equal, $w(T) = \|T\|$. Recall by Theorem 1.3-2 that this condition implies $r(T) = \|T\|$.

An operator is *convexoid* if the closure of its numerical range is the convex hull of its spectrum, $\overline{W(T)} = \operatorname{Co}\sigma(T)$. The notations $\Sigma(T) = \operatorname{co}\sigma(T)$ are also commonly used.

An operator is *spectraloid* (also sometimes called spectral) if its numerical and spectral radii are equal, $r(T) = w(T)$. Note that this means that $\sigma(T) \cap \partial W(T) \neq \phi$.

In this chapter, we study some of the relations between these classes and see what additional conditions, especially those related to the numerical range, make them normal. There are a number of related operator classes which we do not treat.

6.1 Resolvent Growth

There are considerable differences between the three classes of operators mentioned in the Introduction, in spite of the fact that they share some properties of the normal operator. However, additional structure often permits us to establish the equality of any two of these classes. In this section, we provide some technical lemmas concerned with resolvent growth, an underlying theme in this chapter, which enable us to establish such equalities. This is done here to avoid breaking the cycle of arguments presented later on.

Lemma 6.1-1 (Bare points). *A nonempty, compact convex set in $\mathbb{C}$ is the convex hull of its bare points.*

A complex number λ is a *bare point* of a set S in $\mathbb{C}$ if there exists a circle passing through λ containing S. This lemma replaces the extreme points in the Krein–Milman theorem with the bare points; see Notes and References.

Proof. Let S be the compact convex set and Δ the set of its bare points. If S consists of one point, this is in Δ. If S consists of at least two points, its diameter is assumed at two distinct points α and β. It is clear that α and β are in Δ by considering circles centered at one point and passing through the other. Thus Δ is nonempty. Let D be the closed, convex hull of Δ. Evidently, $D \subset S$. Suppose that $D \neq S$. Then, there is a support line L of D so that D lies on one side of L and the point $s \in S - D$ on the other side.

Without loss of generality, let us assume that L is the imaginary axis and D is contained in the left half-plane. Now suppose that the set $E = S \cap \{(x,y) : x \leq 0\}$ is contained in the region

$$\{(x,y) : -b \leq x \leq 0, -M \leq y \leq M\}.$$

This is always possible since S is bounded. The circle with center at $(-c, 0)$ and passing through $(\frac{a}{2}, 0)$ contains E if

$$ac > \sqrt{M^2 + b^2} \ .$$

Thus, the set of circles containing E and having a nonempty intersection with $S - D$ is nonempty. Hence, there exists a circle with center at $(-c, 0)$ and having the smallest radius, which touches S at the point d, with $\operatorname{Re} d > 0$. Then d is a bare point not in D, a contradiction. □

Lemma 6.1-2. *The conditions*

(6.1-1) $$\|(T - \lambda I)^{-1}\| d(\lambda, \sigma(T)) \leq 1$$

and

(6.1-2) $$d(\lambda, \sigma(T)) \|x\| \leq \|Tx - \lambda I x\| \quad \text{for all} \quad x \in H$$

are equivalent for any $\lambda \notin \sigma(T)$.

Proof.
$$\|(T - \lambda I)^{-1}\| d(\lambda, \sigma(T)) \leq 1 \iff$$
$$\|(T - \lambda I)^{-1} y\| d(\lambda, \sigma(T)) \leq \|y\| \ textforall \ y \in H \iff$$
$$\|x\| d(\lambda, \sigma(T)) \leq \|Tx - \lambda I x\| \text{ choosing } y = Tx - \lambda x. \ □$$

Lemma 6.1-3. *T and λ in the previous lemma can be replaced by $\alpha T + \beta I$ and $\alpha \lambda + \beta$, where $\alpha, \beta \in \mathbb{C}$.*

Proof. Notice that
$$(\alpha T + \beta I) - (\alpha \lambda + \beta I) = \alpha(T - \lambda I),$$
$$\|\alpha(T - \lambda I)\| = |\alpha|\|T - \lambda I\|,$$
and $\sigma(\alpha T + \beta I) = \alpha \sigma(T) + \beta$. □

Lemma 6.1-4. *The following inequality holds for any* $\lambda \notin \sigma(T)$:
(6.1-3) $$d(\lambda, \sigma(T))\|(T - \lambda I)^{-1}\| \geq 1.$$

Proof. Since the norm of any operator is not less than the spectral radius, we have

(6.1-4) $$\|(T - \lambda I)^{-1}\| \geq r((T - \lambda I)^{-1}) = \sup\{|\alpha|, \alpha \in \sigma(T - \lambda I)^{-1}\}.$$

Since $(T - \lambda I)$ is invertible, we have $0 \notin \sigma((T - \lambda I)^{-1})$. Taking $\alpha = \frac{1}{\mu}$ and using the spectral mapping theorem, we can thus write

$$\|(T - \lambda I)^{-1}\| \geq \sup\left\{\frac{1}{|\mu|}, \mu \in \sigma(T - \lambda I)\right\}$$

$$= \frac{1}{\inf\{|\mu|, \mu \in \sigma(T - \lambda I)\}}$$

$$= \frac{1}{d(\lambda, \sigma(T))}. \quad \Box$$

Next, we give a condition for the positivity of T, actually, its dissipativeness.

Lemma 6.1-5. *The conditions*

(6.1-5) $$\lambda\|(T - \lambda I)^{-1}\| \leq 1 \quad \text{for all positive} \quad \lambda \notin \sigma(T)$$

and

(6.1-6) $$\operatorname{Re}\langle Tx, x\rangle \leq 0 \quad \text{for all} \quad x \in H$$

are equivalent.

Proof. Let $\lambda\|(T - \lambda I)^{-1}\| \leq 1$. For any $y \in H$, we have $\lambda\|(T - \lambda I)^{-1}y\| \leq \|y\|$. Choosing $y = Tx - \lambda I x$, we get $\lambda\|x\| \leq \|Tx - \lambda I x\|$. Taking $\|x\| = 1$, we have

$$\lambda^2 \leq \|Tx\|^2 - 2\lambda \operatorname{Re}\langle Tx, x\rangle + \lambda^2.$$

Thus $2\lambda \operatorname{Re}\langle Tx, x\rangle \leq \|Tx\|^2$.

Since λ is arbitrary, we must have

$$\operatorname{Re}\langle Tx, x\rangle \leq 0.$$

The converse can be proven by simply reversing the steps. □

Notes and References for Section 6.1

A compact convex set can be obtained by intersecting either the planes containing it or all the circles containing it. This distinction is reflected in considering it as the convex hull of its extreme points or bare points. This idea is used in the improvement of the Krein–Milman theorem, Lemma 6.1-1, by

G. Orland (1964). "On a Class of Operators," *Proc. Amer. Math. Soc.* **15**, 75–79

where the connection between the resolvent norm and the negativity of the operator, Lemma 6.1-5, is also established.

Some more results involving bare points in convexoidity can be found in V. Istratescu and I. Istratescu (1970). "On Bare and Semibare Points for Same Classes of Operators," *Portugaliae Mathematica* **29**, Fax 4, 205–211.

Growth conditions on the resolvent were studied by, among others, G. R. Luecke (1971). "A Class of Operators on Hilbert Space," *Pacific J. Math.* **41**, 153–156.

Many relations between resolvent growth conditions and particular operator classes were given by

T. Furuta (1977). "Relations between Generalized Growth Conditions and Several Classes of Convexoid Operators," *Canadian J. Math.* **29**, 1010–1030.

Lemmas 6.11-2 and 6.1-4 are essentially the general relation (1.4-6), which we stated earlier. For $\lambda \notin \overline{W(T)}$, this general relationship becomes (4.6-7), which we also stated earlier.

6.2 Three Classes

The normaloid operators have a simple characterization in terms of the norms of powers of T.

Theorem 6.2-1. *An operator T is normaloid iff*

(6.2-1) $$\|T^n\| = \|T\|^n \quad \text{for} \quad n = 1, 2, 3, \ldots .$$

Proof. Let T be normaloid. Then $\|T^n\| = [r(T)]^n = r(T^n)$ by the spectral mapping theorem and hence

$$\|T\|^n = [r(T)]^n \leq \|T^n\|.$$

Since we always have

$$\|T\|^n \geq \|T^n\|$$

we conclude that $\|T^n\| = \|T\|^n$.

On the other hand, if
$$\|T^n\| = \|T\|^n \quad \text{for} \quad n = 1, 2, 3, \ldots,$$
we have
$$r(T) = \lim_{n \to \infty} \|T^n\|^{1/n} = \lim_{n \to \infty} \|T\| = \|T\|. \quad \square$$

The following cycle of theorems relates the normaloid and convexoid operators.

Theorem 6.2-2. *An operator T is convexoid if $T - \lambda I$ is normaloid for all $\lambda \in \mathbb{C}$.*

Proof. Let $T - \lambda I$ be normaloid for all $\lambda \in \mathbb{C}$. In view of Lemma 6.1-1, it is enough to prove that every bare point $\mu \in \overline{W(T)}$ is in $\sigma(T)$. There exists a circle through μ and containing $\overline{W(T)}$. Let λ be the center of this circle. Then
$$|\mu - \lambda| = \sup\{d(\lambda, z) : z \in \overline{W(T)}\}$$
$$= \sup\{d(0, z - \lambda), z \in \overline{W(T)}\}$$
$$= w(T - \lambda I) = \|T - \lambda I\|$$
since $T - \lambda I$ is normaloid. Since $\mu - \lambda \in \overline{W(T - \lambda)}$ and $|\mu - \lambda| = \|T - \lambda I\|$, we have $\mu - \lambda \in \sigma(T - \lambda I)$. Hence $\mu \in \sigma(T)$. $\square$

Theorem 6.2-3. *If T is convexoid, then*
$$\|(T - \lambda I)^{-1}\| d(\lambda, \Sigma(T)) \leq 1 \quad \text{for all} \quad \lambda \notin \Sigma(T).$$

Proof. Since $\Sigma(T)$ is a compact convex set, given $\lambda \in \mathbb{C}$, there exists a point μ on $\Sigma(T)$ so that $d(\lambda, \mu) = d(\lambda, \Sigma(T))$. In view of Lemma 6.1-3, we can choose μ as the origin and the line joining λ and μ as the x-axis. Since T is convexoid, we may assure that $W(T)$ also lies in the left half plane and hence $\operatorname{Re} T \leq 0$. Then using Lemma 6.1-5, we have
$$\|(T - \lambda I)^{-1}\| \leq \frac{1}{\lambda} = \frac{1}{d(\lambda, \Sigma(T))}. \quad \square$$

Theorem 6.2-4. *If an operator T satisfies the inequality*
$$\|(T - \lambda I)^{-1}\| d(\lambda, \Sigma(T)) \leq 1$$
for all $\lambda \notin \Sigma(T)$, then T is convexoid.

Proof. Without loss of generality, in view of Lemma 6.1-3, we may assume that $\Sigma(T)$ has the imaginary axis as a line of support and lies in the left half-plane. Let λ be any point on the real axis with $\lambda > 0$. We then have
$$\|(T - \lambda I)^{-1}\| \leq \frac{1}{\lambda},$$

implying, as in Lemma 6.1-5, that $\operatorname{Re} T \leq 0$. Hence $W(T)$ is also in the left half-plane. □

Summarizing Theorems 6.2-2, 6.2-3, and 6.2-4, we have

(6.2-2)
$$T - \lambda I \text{ is normaloid for all } \lambda \in \mathbb{C} \Longrightarrow$$
$$T \text{ is convexoid } \iff \|(T - \lambda I)^{-1}\| d(\lambda, \Sigma(T)) \leq 1$$

for all $\lambda \notin \Sigma(T)$.

The next two theorems relate and describe convexoid and spectraloid operators. It is obvious from the definitions that convexoid operators are spectraloid. In the spirit of Theorem 6.2-2, we can see that if the translates of T are spectraloid, then T is convexoid.

Theorem 6.2-5. *An operator T is convexoid iff $T - \lambda I$ is spectraloid for all $\lambda \in \mathbb{C}$.*

Proof. If T is convexoid, so is $T - \lambda I$ and hence spectraloid. To prove the converse, let us observe that any compact convex set X in $\mathbb{C}$ can be written as the intersection of all the circles containing it. Thus

(6.2-3)
$$\overline{W(T)} = \bigcap_\alpha \left\{ \lambda : |\lambda - \alpha| \leq \sup_{x \in \overline{W(T)}} |x - \alpha| \right\}$$
$$= \bigcap_\alpha \{\lambda : |\lambda - \alpha| \leq w(T - \alpha I)\}.$$

Similarly,

$$\Sigma(T) = \bigcap_\alpha \{\lambda : |\lambda - \alpha| \leq r(T - \alpha I)\}.$$

Since $T - \alpha I$ is spectraloid, we observe that the above sets are identical. □

Theorem 6.2-6 (Power equality). *An operator T is spectraloid iff*

(6.2-4) $\qquad w(T^k) = w^k(T), \quad \text{for} \quad k = 1, 2, 3, \ldots .$

Proof. Assuming that

$$w(T^k) = w^k(T),$$

we have

$$w^k(T) \leq \|T^k\|.$$

Hence $w(T) \leq \|T^k\|^{1/k}$. Taking the limit as $k \to \infty$, we have

$$w(T) \leq r(T).$$

The reverse inequality always holds.

To prove the converse, let $\lambda \in \sigma(T)$ such that $|\lambda| = r(T) = w(T)$. By the spectral mapping theorem, we have

$$\lambda^n \in \sigma(T^n) \subset W(T^n),$$

and hence

$$|\lambda|^n = w^n(T) \leq w(T^n).$$

The reverse inequality $w(T^n) \leq w^n(T)$ is the power inequality theorem (Theorem 2.1-1). $\square$

Example 1. A well-known class of normaloid operators is that of paranormal operators. An operator is *paranormal* if

(6.2-5) $\qquad \|Tx\|^2 \leq \|T^2x\|\|x\| \quad \text{for all} \quad x \in H.$

Let us show that paranormal operators are always normaloid. Taking $\|x\| = 1$, we have

$$\|Tx\|^2 \leq \|T^2x\| \leq \|T^2\| \quad \text{for all} \quad \|x\| = 1.$$

Hence $\|T\|^2 \leq \|T^2\|$. Since we always have $\|T^2\| \leq \|T\|^2$, we conclude that $\|T\|^2 = \|T^2\|$. An induction argument shows that $\|T^n\| = \|T\|^n$. We can now use Theorem 6.2-1 to conclude that T is normaloid.

Paranormal operators are not always convexoid (see Notes and References). However, the hyponormal operators, a subclass of the paranormal operators, are convexoid. An operator T is *hyponormal* if

(6.2-6) $\qquad \|T^*x\| \leq \|Tx\| \quad \text{for all} \quad x \in H.$

We see immediately that

$$\|Tx\|^2 = \langle T^*Tx, x \rangle \leq \|T^*Tx\|\|x\| \leq \|T^2x\|\|x\|,$$

and hence the hyponormal operators are also paranormal and thus normaloid. It is easy to check that all translates $T - \lambda I$ of a hyponormal operator T are also hyponormal. Hence T is convexoid using Theorem 6.2-2.

Example 2. A normaloid operator need not be convexoid. Let $H = \mathbb{C}^3$ with the Euclidean norm, given by

$$\|f\|^2 = \|(f_1, f_2, f_3)\|^2 = |f_1|^2 + |f_2|^2 + |f_3|^2.$$

Let

$$T = \begin{bmatrix} 0 & 1 & 0 \\ 0 & 0 & 0 \\ 0 & 0 & 1 \end{bmatrix}.$$

Then $Tf = (f_2, 0, f_3)$ and $\|T\|^2 = \sup_{\|f\|=1}\{|f_2|^2 + |f_3|^2\} = 1$. On the other hand,

$$\langle Tf, f \rangle = f_2 \bar{f}_1 + f_3 \bar{f}_3,$$

and consequently,
$$w(T) = \sup_{\|f\|}\{f_2\bar{f}_1 + f_3\bar{f}_3\} = 1$$
by taking $f_3 = (0,0,1)$. Hence T is normaloid.

It can be verified easily that
$$\sigma(T) = \{(1,0)\}$$
and $W(T)$ is the smallest convex set containing $(1,0)$ and the set
$$\left\{\lambda, |\lambda| < \frac{1}{2}\right\}.$$
Hence T is not convexoid.

Example 3. A convexoid operator need not be normaloid. Let $\{x_1, x_2, \ldots\}$ be an orthonormal base for $H = \ell_2$. Define $z_n = x_{2n+1}$, $n = 0, 1, 2, \ldots$, and $z_{-n} = x_{2n}$, $n = 1, 2, 3, \ldots$. Every $x \in H$ can be written as
$$x = \sum_{-\infty}^{\infty} \alpha_k z_k.$$
Let us now define the operator S on H by
$$Sx = \frac{1}{2}\sum_{-\infty}^{\infty} \alpha_k z_{k+1},$$
where $x = \sum_{-\infty}^{\infty} \alpha_k z_k$. We can check easily that
$$W(S) = \left\{\lambda \in \mathbb{C}, |\lambda| \leq \frac{1}{2}\right\}.$$
Let us define the operator
$$L = \begin{bmatrix} 0 & 1 \\ 0 & 0 \end{bmatrix} \quad \text{on} \quad \mathbb{C}^2.$$
The operator T defined on $H \oplus \mathbb{C}^2$ by
$$T(f, g) = (Lf, Sg)$$
yields
$$W(T) = \left\{\lambda \mathbb{C}, |\lambda| < \frac{1}{2}\right\} = \mathrm{co}\,\sigma(T).$$
T is not normaloid since
$$\|T\| = 1 \quad \text{and} \quad w(T) = \frac{1}{2}.$$

Example 4. A spectraloid operator need not be convexoid. In $H = \mathbb{C}^3$, consider the operator

$$T = \begin{bmatrix} 1 & 0 & 0 \\ 0 & 0 & 0 \\ 0 & 1 & 0 \end{bmatrix}.$$

We have, as in Example 2, that $\sigma(T) = \{0, 1\}$ and $W(T) = \mathrm{co}\,\{\sigma(T), S\}$, where

$$S = \left\{\lambda \in \mathbb{C}, |\lambda| \leq \frac{1}{2}\right\}.$$

Example 5. A slight modification of the above example produces a spectraloid operator that is not normaloid. In $H = \mathbb{C}^3$, let the operator

$$T = \begin{bmatrix} 1 & 0 & 0 \\ 0 & 0 & 0 \\ 0 & 2 & 0 \end{bmatrix}.$$

We have $\|T\| = 2$ and $w(T) = r(T) = 1$.

Notes and References for Section 6.2

Normaloid operators were first studied by
A. Wintner (1929). "Zur theorie beschränkten Bilinearformem," *Math. Z.*, Vol. 30, 228–282.

The nomenclature "convexoid" and "spectraloid" is due to Halmos [H], where several examples of proper inclusion between these classes are provided.

An exhaustive characterization of convexoid operators can be found in
T. Furuta (1973). "Some Characterizations of Convexoid Operators," *Rev. Roum. Math. Pures et Appl.* **18**, 893–900.

The same author raised the interesting question of whether paranormal operators are always convexoid, in
T. Furuta (1967). "On the Class of Paranormal Operators," *Proc. Japan Acad.* **43**, 594–598.

Counterexamples were given independently in
T. Ando (unpublished, 1974).
K. Gustafson and D. Rao (unpublished, 1985).
For the latter, see
D. K. Rao (1987). "Operadores Paranormales," *Revista Colombiana de Matematicas* **21**, 135–149.

The original proof for the convexoidity of hyponormal operators was given by
J. G. Stampfli (1965). "Hyponormal Operators and Spectral Density," *Trans. Amer. Math. Soc.* **117**, 469–476.

The condition $\|(T - \lambda I)^{-1}\| \leq (d(\lambda, \Sigma(T)))^{-1}$ (without the use of the name convexoid) gave rise to a corresponding class of operators in the early 1960s. Some of the results proved in Orland's paper (see References for Section 6.1) also appear in

G. Lumer (1961). "Semi-inner-product Spaces," *Trans. Amer. Math. Soc.* **100**, 24–43.

G. Lumer and R. S. Phillips (1961). "Dissipative Operators in a Banach Space," *Pacific J. Math.* **11**, 679–698.

in the context of operators on Banach spaces rather than Hilbert spaces.

In Section 1.4, we commented on the fact that normal operators satisfied the exact estimate (1.4-4), namely, that $\|(T - \lambda I)^{-1}\|^{-1} = d(\lambda, \sigma(T))$ for all $\lambda \notin \sigma(T)$. This condition was one of the early jumping-off points in the development of the operator classes discussed in this section. The class of all operators T satisfying this condition was called the *class G_1*. It turns out that this class falls properly between the hyponormal operators and the convexoid operators.

Another class of operators, known as *centroid operators*, was introduced in

S. Prasanna (1981). "The Norm of a Derivation and the Björck–Thomee–Istratescu Theorem," *Math Japonica* **26**, 585–588.

Let D_σ and D_W be the smallest disks containing $\sigma(T)$ and $W(T)$, respectively, with radii R_T and W_T and centers at z_T and w_T. Further, let $B_T = \sup_{\|x\|=1}\{\|Tx\|^2 - |\langle Tx, x\rangle|^2\}$. Then $M_T = \sqrt{B_T}$ is the distance of T to the scalars and is called the *transcendental radius* of T. The operator T is called centroid if $T - Z_T$ is normaloid. Clearly, centroid operators are a class of operators beyond spectraloid. For further information, see

T. Furuta, S. Izumino and S. Prasanna (1982). "A Characterization of Centroid Operators," *Math. Japonica* **27**, 105–106.

6.3 Spectral Sets

A set S is called *spectral* for an operator T if it contains $\sigma(T)$ and

(6.3-1) $$\|f(T)\| \leq \sup\{|f(z)|, z \in S\}$$

for all rational functions f with no poles in S.

Notice that if S is a spectral set for T, any set containing S will also be spectral.

Theorem 6.3-1. *Let $\overline{W(T)}$ be spectral for T. Then $T - \lambda I$ is normaloid for all $\lambda \in \mathbb{C}$.*

Proof. Consider

$$f(z) = z - \lambda.$$

Then, the spectrality of $\overline{W(T)}$ implies

(6.3-2)
$$\|T - \lambda I\| \leq \sup\{|z - \lambda|,\ z \in \overline{W(T)}\}$$
$$= d(\lambda, \overline{W(T)}) = w(T - \lambda I).$$

Hence $T - \lambda I$ is normaloid. □

Corollary 6.3-2. If $\overline{W(T)}$ is a spectral set for T, then T is convexoid.

Proof. This follows immediately from Theorem 6.2-2. □

Corollary 6.3-3. If $\operatorname{co} \sigma(T)$ is a spectral set for T, then $T - \lambda I$ is normaloid for all $\lambda \in \mathbb{C}$ and T is convexoid.

Proof. If $\operatorname{co} \sigma(T)$ is spectral for T, so is $\overline{W(T)}$. □

We now characterize the operators T for which $\overline{W(T)}$ is a spectral set. We borrow a theorem from dilation theory (see Notes and References) which says that for any compact convex set S that is spectral for T, there exists a normal operator N defined on a larger Hilbert space $K \supset H$ such that $\sigma(N) \subset \partial S$ and $T^n x = PN^n x$, $x \in H$, $n = 0, 1, 2, \ldots$. Such a dilation N is called a strong normal dilation of T (see Section 2.6).

Theorem 6.3-4. If $\operatorname{co} \sigma(T)$ is spectral for T, then there exists a strong normal dilation N of T such that

(6.3-3)
$$\operatorname{co} \sigma(T) = \overline{W(T)} = \overline{W(N)}.$$

Proof. Let N be the dilation such that $\sigma(N) \subset \partial \operatorname{co} \sigma(T)$ and $T^n x = PN^n x$ for $x \in H$, $n = 0, 1, 2, \ldots$. Then $\overline{W(T)} = \{\langle Nx, x \rangle \mid x \in H, \|x\| = 1\} \subset \overline{W(N)}$. Since N is normal, $\overline{W(N)} = \operatorname{co} \sigma(N)$. Hence $\overline{W(T)} \subset \operatorname{co} \sigma(N) = \overline{W(N)}$. Also, $\operatorname{co} \sigma(N) \subset \operatorname{Co} \partial \operatorname{Co} \sigma(T) = \operatorname{Co} \sigma(T) \subset \overline{W(T)}$. So $\operatorname{co} \sigma(T) = \overline{W(T)} = \overline{W(N)}$. □

From Theorem 6.3-4, one immediately obtains the following corollary.

Corollary 6.3-5. If $\overline{W(T)}$ is a spectral set for T, then there is a strong normal dilation N of T such that $\overline{W(T)} = \overline{W(N)}$.

Proof. By Corollary 6.3-2, T is convexoid. □

Theorem 6.3-6. If there is a strong normal dilation N of T such that $\overline{W(T)} = \overline{W(N)}$, then $\overline{W(T)}$ is a spectral set for T.

Proof. For any λ, $|\lambda| > \|N\| = w(N) = w(T)$, we have series expansions for $(T - \lambda I)^{-1}$ and $(N - \lambda I)^{-1}$, where $T^n = PN^n$. So $\langle (T - \lambda I)^{-1} x, y \rangle =$

$\langle (N-\lambda I)^{-1}x, y\rangle$ for $x, y \in H$. Since this equality holds for all $\lambda \notin \overline{W(T)}$, we have for any rational function f that has no poles in $\overline{W(T)}$,

$$\langle f(T)x, y\rangle = \langle f(N)x, y\rangle, \quad x, y \in H.$$

Hence

$$\|f(T)\| \leq \|f(N)\| = \sup\{|f(z)|, z \in \sigma(N)\}$$
$$\leq \sup\{|f(z)|, z \in \overline{W(N)} = \overline{W(T)}\}. \qquad \square$$

Notes and References for Section 6.3

Spectral sets were first introduced by Von Neumann in an attempt to extend spectral theory to nonnormal operators. For polynomials $p(T)$ of a contraction ($\|T\| \leq 1$) and S the closed unit disk, the spectral set condition (6.3-1) is satisfied. See
J. von Neumann (1951). "Eine Spektraltheorie für Allgemeine Operatoren eines Unitären Raumes," *Math. Nachr.* **4**, 258–281.

See [RN] for the classical theory of spectral sets. Their relation to numerical range was studied by
M. Schreiber (1963). "Numerical Range and Spectral Sets," *Michigan Math. J.* **10**, 283–288.
S. Hildebrandt (1964). "The Closure of the Numerical Range as a Spectral Set," *Comm. Pure Appl. Math.*, 415–421.
See also
S. Hildebrandt (1966). "Über den Numerischen Wertebereich eines Operators," *Math. Annalen* **163**, 230–247.
K. Gustafson (1972). "Necessary and Sufficient Conditions for Weyl's Theorem," *Michigan Math. J.* **19**, 71–81.

The theorem we used for dilation is due to
A. Lebow (1963). "On Von Neumann's Theory of Spectral Sets," *J. Math. Anal. Appl.* **7**, 64–90.

The simple proof of Theorem 6.3-1 is due to
T. Saito and T. Yoshino (1965). "On a Conjecture of Berberian," *Tohoku Math. J.* **17**, 147–149.

If a triangle contains $W(T)$, then the circumcircle of this triangle is a spectral set for T because at least one of the vertices is at a distance from the origin greater than or equal to $\|T\|$. This result is given in
B. A. Mirman (1968). "Numerical Range and Norm of a Linear Operator," *Trudy Seminara po Funkcional Analizu* **10**, 51–55.
From this it follows that if the numerical range $W(T)$ is a triangle, then for all complex numbers α and β, the operator $\alpha T + \beta I$ is normaloid.

6.4 Normality Conditions

The three classes of operators described in Section 6.2 are far from being normal, although they share some properties with normal operators. Now we will present some results relating these classes to normality.

Theorem 6.4-1. *The following conditions on an operator T are equivalent:*

(6.4-1)
    (a) T is normal.
    (b) Every quadratic polynomial in T and T^* is normaloid.
    (c) Every quadratic polynomial in T and T^* is convexoid.
    (d) Every quadratic polynomial in T and T^* is spectraloid.

Proof. $(a) \implies (b)$. If T is normal, so is every quadratic polynomial $p(T, T^*)$ in T and T^*. Consequently, $p(T, T^*)$ is normaloid.

$(b) \implies (c)$. The quadratic polynomials $p(T, T^*) + \lambda I$, $\lambda \in \mathbb{C}$ are all normaloid, and hence $p(T, T^*)$ is convexoid by Theorem 6.2-2.

$(c) \implies (d)$. The family of quadratic polynomials $p(T, T^*)$ is also spectraloid.

$(d) \implies (c)$. The family of quadratic polynomials $p(T, T^*) + \lambda I$, $\lambda \in \mathbb{C}$ are all spectraloid and hence $p(T, T^*)$ is convexoid, using Theorem 6.2-5.

$(c) \implies (a)$. Let A and B be the selfadjoint operators given by

$$A = \frac{1}{2}(T + T^*) = \operatorname{Re} T,$$

$$B = \frac{1}{2i}(T - T^*) = \operatorname{Im} T.$$

Suppose that the quadratic polynomial AB,

$$AB = \frac{1}{4i}(T - T^*)(T + T^*),$$

is convexoid. Without loss of generality, we may assume that A and B are strictly positive. Then we have, using Theorem 2.4-1,

(6.4-2) $$\sigma(AB) \subset \frac{W(B)}{W(A^{-1})}.$$

Hence $\sigma(AB)$ is real. Since AB is convexoid, we see that $W(AB)$ is real. Hence AB is selfadjoint, and this implies the normality of T. $\square$

Other conditions equivalent to the normality of an operator T in a finite-dimensional Hilbert space depend on the notion of reduction properties. For any property P of operators, we say that an operator T is *reduction-P* if the restriction of T to any reducing (invariant under T and T^*) subspace has property P. Similarly, an operator T is *restriction-P* if its restriction to every invariant subspace has property P. Further, let us denote an operator T as *transloid* if $\alpha T + \beta I$ is normaloid for $\alpha, \beta \in \mathbb{C}$.

Theorem 6.4-2. *The following conditions on an operator T in a finite-dimensional Hilbert space are equivalent:*

(6.4-3)
- (a) T is normal.
- (b) T is reduction transloid.
- (c) T is reduction normaloid.
- (d) T is reduction convexoid.
- (e) T is reduction spectraloid.

Proof. The proof depends on the concept of normal eigenvalue introduced in Theorem 5.1-9, namely, an eigenvalue on the boundary of the numerical range. Recall that if λ is a normal eigenvalue, then the dimension of the eigenspace M for λ is equal to its algebraic multiplicity and A is unitarily equivalent to $\lambda I_M \oplus B$, where $\lambda \notin \sigma(B)$.

If T is normal, so is its restriction to any reducing subspace, and so condition (a) implies (b), (c), and (d). Also, (e) and (b) $\Rightarrow$ (c) by definition; (b) $\Rightarrow$ (d) by Theorem 6.2-2; (d) $\Rightarrow$ (e) and (c) $\Rightarrow$ (e) by definition. So the only nontrivial implication remaining to be shown is (e) $\Rightarrow$ (a).

Let T be spectraloid on a reducing subspace M. Choose $\lambda \in \sigma(T/M)$ with $|\lambda| = w(T/M)$. Then λ is a normal eigenvalue for T, and using Theorem 5.1-9 we can write $T = \lambda I_n \oplus B$, where I_N is the eigenspace for λ. We can now deal with B in a finite number of similar steps. □

Notes and References for Section 6.4

Conditions for normality of the classes of operators considered were given by
S. K. Berberian (1970). "Some Conditions on an Operator Implying Normality," *Math. Ann.* **184**, 188–192.

Earlier, using arbitrary polynomials instead of quadratic polynomials, a condition of normality was obtained by
R. G. Douglas and P. Rosenthal (1968). "A Necessary and Sufficient Condition for Normality," *J. Math. Anal. Appl.* **22**, 10–11.

Earlier results can be found in
J. G. Stampfli (1962). "Hyponormal Operators," *Pacific J. Math.* **12**, 1453–1458.

Further conditions on convexoid operators to attain normality are given in
C. Meng (1963). "On the Numerical Range of an Operator," *Proc. Amer. Math. Soc.* **14**, 167–171.

Additional conditions equivalent to normality using the restriction-P concept can be seen in the paper by Berberian (1970). A restriction-normaloid operator T is normal if it is compact, as shown by V. Istratescu and I. Istratescu (1967). "On Normaloid Operators," *Math. Zeitschr.* **105**, 153–156.

Conditions equivalent to normality of arbitrary matrices were given by R. Grone, C. Johnson, E. Sa and H. Wolkowicz (1987). "Normal Matrices," *Lin. Alg. and Appl.* **87**, 213–225.

6.5 Finite Inclusions

Additional relations between the three basic classes of normal-like operators that we have discussed can be obtained for certain finite dimensions. We will study such relations principally for dimensions $n = 2, 3, 4$.

Following the notation of common use in this literature, let us denote, for general n,

(6.5-1)
$$\begin{aligned}
N_n &= \text{normal operators } \in M_n, \\
R_n &= \text{normaloid operators } \in M_n, \\
Q_n &= \text{convexoid operators } \in M_n, \\
S_n &= \text{spectraloid operators } \in M_n.
\end{aligned}$$

We always have $S_n \subset R_n$, $N_n \subset R_n \cap Q_n \cap S_n$, $Q_n \subset S_n$.

Theorem 6.5-1. $N_2 = R_2 = S_2$.

Proof. It is enough to show $S_2 \subset N_2$. By Schur's lemma, we may assume that
$$T = \begin{bmatrix} \lambda_1 & a \\ 0 & \lambda_2 \end{bmatrix},$$
where $\lambda_1, \lambda_2 \in \sigma(T)$ and $|\lambda_1| = r(T) = w(T)$, the latter since T is spectraloid. But if $a \neq 0$, λ_1 is a focus in the interior of an ellipse, a contradiction. If $a \equiv 0$, T is diagonalizable. $\square$

The argument used above can be used to characterize spectraloid operators in M_n.

Theorem 6.5-2. $T \subset S_n$ iff T is unitarily similar to a matrix of the form
$$T = \begin{bmatrix} A & 0 \\ 0 & B \end{bmatrix},$$
where A is diagonal and B is triangular, with $w(B) \leq r(T)$.

Proof. We can assume that T is triangular and its eigenvalues $\lambda_1, \lambda_2, \ldots$ are ordered as
$$|\lambda_1| = |\lambda_2| = \cdots = |\lambda_s| \geq |\lambda_{s+1}| \geq \cdots \geq |\lambda_n|, \quad s \geq 1.$$
Here s is the number of eigenvalues whose magnitude is equal to the numerical radius of T. Let us choose i, j such that $1 \leq i \leq s$ and $s + 1 \leq j \leq n$. Consider the 2×2 submatrix of T
$$S = \begin{bmatrix} \lambda_i & t_{ij} \\ 0 & \lambda_j \end{bmatrix},$$
where $|\lambda_i| \geq |\lambda_j|$. By the submatrix inclusion $W(S) \subset W(T)$, $w(S) \leq w(T) = |\lambda_i|$. As in Theorem 6.5-1, we can show $w(S) > |\lambda_i|$ unless $t_{ij} = 0$, $r(T) = w(T) = \max\{w(A), w(B)\} = |\lambda_1|$. The proof in the other direction is straightforward. □

Corollary 6.5-3. $N_2 = Q_2 = R_2 = S_2$.

Proof. Observe that $Q_2 \subset S_2$. □

Theorem 6.5-4. $N_3 = Q_3$.

Proof. If the three eigenvalues $\lambda_1, \lambda_2, \lambda_3$ of $T \in Q_3$ are collinear, then T is normal since a rotation of a translate of T is selfadjoint.

Let us then consider the general case when $W(T)$ is the triangle $\triangle$ whose vertices are $\lambda_1, \lambda_2, \lambda_3$. T can be assumed to be
$$T = \begin{bmatrix} \lambda_1 & t_{12} & t_{13} \\ 0 & \lambda_2 & t_{23} \\ 0 & 0 & \lambda_3 \end{bmatrix}.$$
Consider the submatrix
$$S = \begin{bmatrix} \lambda_1 & t_{13} \\ 0 & \lambda_3 \end{bmatrix}.$$
$W(S)$ is an ellipse with foci at λ_1, λ_3 and minor axis $|T_{13}|$. This ellipse is not contained in $\triangle$ unless $t_{13} = 0$. Since $T \in Q_3$, t_{13} must be zero. Similarly, $t_{12} = t_{23} = 0$. Hence T is diagonalizable. □

Corollary 6.5-5. $N_4 = Q_4$.

Proof. If $\lambda_1, \lambda_2, \lambda_3, \lambda_4$ form a quadrilateral, we can use the argument of Theorem 6.5-4. There can be at most one point that is not a vertex of the triangle formed by the three other eigenvalues, and the same argument carries through. □

Corollary 6.5-6. $Q_4 = R_4$.

Proof. $Q_4 = N_4$. □

Summarizing the above results, we have the following situation.

(6.5-2)
$$N_2 = Q_2 = R_2 = S_2, \quad n = 2$$
$$N_n = Q_n \subset R_n \subset S_n, \quad n = 3, 4.$$

Some of the proper inclusions can be seen in the examples of Section 6.2.

Notes and References for Section 6.5

Basic theorems useful in this context were given by
B. N. Moyls nd M. D. Marcus (1955). "Field Convexity of a Square Matrix," *Proc. Amer. Math. Soc.* **6**, 981–983.

An extensive study of these relations can be found in
M. Goldberg and G. Zwas (1974). "On Matrices Having Equal Spectral Radius and Spectral Norm," *Lin. Alg. Appl.* **8**, 427–434.
M. Goldberg, E. Tadmore and G. Zwas (1975). "The Numerical Radius and Spectral Matrices," *Lin. Multilin. Alg.* **2**, 317–326.
M. Goldberg and G. Zwas (1976). "Inclusion Relations Between Certain Sets of Matrices," *Lin. Multilin. Alg.* **4**, 55–60.

6.6 Beyond Spectraloid

From the viewpoint of operator theory, classes of operators withnormal-like properties, such as the normaloid, convexoid, and spectraloid operators, are interesting structurally within the class of all normal-like operators. For example, the *subnormal* operators enlarged the class of operators possessing nontrivial, closed invariant subspaces (see [H]). However, we would like to advance the perspective in this final section that a substantial part of the future of numerical range research must reach beyond the normal-like operators.

For example, consider the operators T and S used in the counter-example to the double-commute conjecture that was discussed in Section 2.5 and was also mentioned in Section 5.3. These are operators T and S with known positive $w(T)$ and $w(S)$ and for which $w(TS)$ is greater than $w(T)\|S\|$. On the other hand, the spectral radii $r(S) = r(T) = r(TS)$ are all zero. In other words, these rather simple finite composite shift operators, all of which would be spectraloid in their infinite-dimensional versions, are already well beyond normal-like in finite dimensions.

Another example of this type is

(6.6-1)
$$S = \begin{bmatrix} S_3 & 0 & 0 \\ 0 & S_3 & 0 \\ 0 & 0 & S_3 \end{bmatrix}, \quad S_3 = \begin{bmatrix} 0 & 1 & 0 \\ 0 & 0 & 1 \\ 0 & 0 & 0 \end{bmatrix}.$$

6.6 Beyond Spectraloid

Forming

$$T = \begin{bmatrix} 0 & I_3 & S_3 \\ 0 & 0 & I_3 \\ 0 & 0 & 0 \end{bmatrix}, \quad I_3 = \text{Identity},$$

we have $ST = TS$, $\|S\| = 1$. For any $x = (x_1, \ldots, x_9)$,

$$Tx = (x_4 + x_8, x_5 + x_9, x_6, x_7, x_8, x_9, 0, 0, 0),$$

and hence

(6.6-2) $\langle Tx, x \rangle = x_4 \bar{x}_1 + x_8 \bar{x}_1 + x_5 \bar{x}_2 + x_9 \bar{x}_2 + x_6 \bar{x}_3 + x_7 \bar{x}_4 + x_8 \bar{x}_5 + x_9 \bar{x}_6.$

As $w(T)$ is the supremum over all expressions in (6.6-2) for $\sum (x_i)^2 = 1$ and $x_i \geq 0$, we may permute T by mapping as follows:

$$(x_1, \ldots, x_9) \to (y_7, y_4, y_1, y_8, y_5, y_2, y_9, y_6, y_3).$$

Then

(6.6-3) $$w(T) = \sup_{\Sigma y_i^2 = 1} \sum_{i=1}^{8} y_i y_{i+1} = \cos\left(\frac{\pi}{10}\right),$$

so that T, composed from S_6 and S_2 submatrices, has the same numerical radius as S_9. One can show that ST has a cycle involving vertices 1, 5, and 9, so $w(ST) = 1$.

Finite shifts are essential in many parts of electrical engineering and in finite quantum field models. As $n \to \infty$, there is the significant spectral radius discontinuity from $r(S_n) = 0$ to $r(S_\infty) = 1$. Moreover, perturbed finite shifts are commonly used (as we did in Chapter 4, in connection with pseudo-eigenvalue theory) to demonstrate high instability of spectra under tiny perturbations. Yet the good stability properties of the numerical range $W(A)$ remain. As $n \to \infty$, the numerical range $W(S_n)$ is a disk of radius $\cos \pi / (n+1)$ which smoothly approaches the unit disk numerical range of the infinite-dimensional shift operator. In other words, not only does the numerical range $W(A)$ possess good stability properties under additive perturbation, for certain operator classes it may also have good stability properties during dimension change.

We have just spoken in terms of Chapters 2, 4, and 5 about going beyond spectraloid. Recall also that the discretizations of most gas dynamics equations, such as those treated in Chapter 4, are not normal, and as shown there, are not hyponormal, and are not generally expected to be spectraloid. Indeed, if they were spectraloid, the hydrodynamic stability problems would all be solvable in terms of convex hull of the spectra of the operators. Thus, the numerical range emerges as an alternate tool, as the role of the spectra no longer suffices.

In Chapter 3, we noted that we need to better understand antieigenvalues for operators beyond normal. With that theory's fundamental geometrical content and relationship to the numerical range, we may expect that more specifics for antieigenvalues and antieigenvectors for operators beyond spectraloid will eventually materialize. Moreover, regarded as a condition number $\mu(HA)$ for a matrix A preconditioned by an approximate inverse H, the antieigenvalue μ may be expected to play a role similar to that of so-called log norms already in use for large sparse matrices A in the numerical analysis community and in optimization theory.

There is a useful way to conclude this perspective of reaching beyond spectraloid, especially for finite matrices A. All such matrices may be thought of in terms of their Jordan canonical form $A = D+U$ if we imagine that we already know the appropriate bases for such representations. The superdiagonal portion U is just composed of partial shifts such as we have referred to in this section. Thus, the beyond spectraloid operators are not so far away as they might at first seem.

Notes and References for Section 6.6

An exploration of how the antieigenvalue μ plays the role of condition number in numerical linear algebra may be found in
K. Gustafson (1996). "Trigonometric Interpretation of Iterative Methods," *Proc. Conf. on Algebraic Multilevel Iteration Methods with Applications*, (O. Axelsson, ed.), June 13–15, Nijmegen, Netherlands, 23–29.

Endnotes for Chapter 6

A useful device for understanding the detailed spectral and related numerical range properties of specific operator classes is the *state diagram*. Because good expositions of the state diagram method already exist in the literature, we have not included a discussion of it here. For further recent information, see
K. Gustafson (1996). "Operator Spectral States," *Computers Math. Applic.* to appear.
K. Gustafson and D. Rao (1996). "Spectral States of Normal-like Operators," to appear.

An underlying theme of Chapter 6 has been the appearance and use of the resolvent operator $(T - zI)^{-1}$ and estimates for it. Most of these estimates are really still rather crude: just a distance to the spectrum $\sigma(T)$, or if that won't suffice, to the larger set $W(T)$. Such estimates can sometimes be improved by use of the right variation of the numerical range or can motivate new variations of the numerical range such as the M-numerical range described in Chapter 5.

We would like to elaborate this theme by advancing a further perspective on resolvent estimates that may be useful in future developments related to numerical range theory and application. What is a resolvent operator?

In applications to physics (the infinite-dimensional, or operator theoretic case), it often is an integral operator with a Green's function kernel $G(x, y)$. Its numerical range $W((T - zI)^{-1}) = \langle (T-zI)^{-1}f, f \rangle / \langle f, f \rangle$ thus becomes a double integral over a kernel, which weights the function $f(x)\overline{f(y)}$. If we happened to know the full eigenfunction expansionals for the physical problem, then we can express everything (e.g., the Green's kernel G, the resolvent series $(T - zI)^{-1}$, the Rayleigh quotients $W((T - zI)^{-1})$), in terms of them. It is exactly in going beyond normal operators, and more generally beyond spectraloid operators, that relying just on eigenfunctional and spectral information no longer suffices.

The same perspective about resolvents applies when T is a matrix A. Resolvent estimates $(A-zI)^{-1}$ may be thought of in terms of the numerical range $W((A-zI)^{-1})$, that may be thought of as a double sum over a kernel which implements a joint weighting of products $x_i y_j$. Perhaps numerical range theory could better take into account this point of view in providing better resolvent estimates for use in applications.

Finally, we would like to echo the suggestion given in the Endnotes to Chapter 5. Future numerical range research should make a strong commitment to a better understanding of the many new and important operator classes coming out of computational linear algebra and applications. Almost all of these are beyond spectraloid. But each class carries particular structure properties reflecting those of the class of applications and those of the discretization methods employed for simulation. In this way, the numerical range will remain a vital and growing part of operator theory and matrix analysis.

Bibliography

Abdurakmanov, A. A.
[1] *The Geometry of the Hausdorff Domain in Localization Problems for the Spectrum of Arbitrary Matrices*, Math. USSR Sb. **59** (1988), 39–51.

Agler, J.
[1] *Geometric and Topological Properties of the Numerical Range*, Indiana Univ. Math. J. **31** (1982), 766–767.

Akcoglu, M.
[1] *Positive Contractions on L_1-Spaces*, Math. Z. **143** (1975), 1–13.

Ando, T.
[1] *On a Pair of Commutative Contractions*, Acta Sci. Math. **24** (1961), 88–90.
[2] *Structure of Operators with Numerical Radius One,"* Acta. Sci. Mat. (Szeged) **34** (1973), 11–15.

Ando, T., Horn, R. A. and Johnson, C. R.
[1] *The Singular Values of a Hadamard Product: A Basic Inequality*, Lin. Multilin. Alg. **21** (1987), 345–365.

Antoine, J. P. and Gustafson, K.
[1] *Partial Inner Product Spaces and Semi-inner Product Spaces*, Adv. Math. **41** (1981), 281–300.

Antoniou, I. and Gustafson, K.
[1] *From Probabilistic Descriptions to Deterministic Dynamics*, Physica A **197** (1993), 153–166.
[2] *Dilation of Markov Processes to Dynamical Systems*, (1996), preprint.

Asplund, E. and Ptak, V.
[1] *A Minimax Inequality for Operators and a Related Numerical Range*, Acta Math. **126** (1971), 53–62.

Au-Yeung, Y. and Tsing, N.
[1] *A Conjecture of Marcus on the Generalized Numerical Range*, Lin. Multilin. Alg. **14** (1983), 235–239.

Bibliography

Axelsson, O., Lu, H. and Polman, B.
[1] *On the Numerical Radius of Matrices and Its Application to Iterative Solution Methods*, Lin. Multilin. Alg. **37** (1994), 225–238.

Barbu, V.
[1] *Nonlinear Semigroups and Differential Equations in Banach Spaces*, Editura Academie, Bucharest, 1976.

Bauer, F.
[1] *On the Field of Values Subordinate to a Norm*, Numer. Math. **4** (1962), 103–111.
[2] *Fields of Values and Gershgorin Disksa*, Numer. Math. **12** (1968), 91–95.

Bebiano, N. and da Providencia, J.
[1] *Some Geometrical Properties of the Numerical Range of a Normal Matrix*, Lin. Multilin. Alg. **37** (1994), 83–92.

Berberian, S. K.
[1] *Approximate Proper Vectors*, Proc. Amer. Math. Soc. **113** (1962), 111–114.
[2] *The Numerical Range of a Normal Operator*, Duke Math. J. **31** (1964), 479–483.
[3] *Some Conditions on an Operator Implying Normality*, Math. Ann. **184** (1970), 188–192.

Berberian, S. K. and Orland, G. H.
[1] *On the Closure of the Numerical Range of an Operator*, Proc. Amer. Math. Soc. **18** (1967), 499–503.

Berger, C. A.
[1] *A Strange Dilation Theorem*, Amer. Math. Soc. Not. **12** (1965), 590.

Berger, C. A. and Stampfli, J. G.
[1] *Norm Relations and Skew Dilations*, Acta. Sci. Math. (Szeged) **28** (1967a), 191–195.
[2] *Mapping Theorems for the Numerical Range*, Amer. J. Math. **89** (1967b), 1047–1055.

Bhagwat, K. and Subramanian, A.
[1] *Inequalities Between Means of Positive Operators*, Math. Proc. Camb. Phil. Soc. **83** (1978), 393–401.

Bonsall, F. and Duncan, J.
[1] *Numerical Ranges of Operators on Normed Spaces and Elements of Normed Algebras*, Cambridge Univ. Press, Cambridge, 1971.
[2] *Numerical Ranges II*, Cambridge Univ. Press, Cambridge, 1973.
[3] *Numerical Ranges*, in Studies in Mathematics **21**, (R. G. Bartle, ed.), Mathematical Association of America, Washington, DC, 1980, 1–49.

Bouldin, R.
[1] *The Numerical Range of a Product*, J. Math. Anal. Appl. **32** (1970), 459–467.
[2] *The Numerical Range of a Product II*, J. Math. Anal. Appl. **33** (1971), 212–219.

Brezis, H.
[1] *Operateurs Maximaux Monotones et Semi-groups de Contractions dans les Espaces de Hilbert*, North-Holland, Amsterdam, 1973.

Brualdi, R. A. and Mellendorf, S.
[1] *Regions in the Complex Plane Containing the Eigenvalues of a Matrix*, Amer. Math. Mon. **101** (1993), 975–985.

Cornfeld, I. P., Fomin, S. V., and Sinai, Ya G.
[1] *Ergodic Theory*, Springer-Verlag, New York, 1982.

Crabb, M. J.
[1] *The Powers of an Operator of Numerical Radius One*, Mich. Math. J. **18** (1971), 252–256.

Das, K. C.
[1] *Boundary of Numerical Range*, J. Math. Anal. Appl. **60** (1977), 779–780.

Das, K., Mazumdar, S. and Sims, B.
[1] *Restricted Numerical Range and Weak Convergence on the Boundary of the Numerical Range*, J. Math. Phys. Sci. **21** (1987), 35–41.

Davidson, K. R. and Holbrook, J. A. R.
[1] *Numerical Radii of Zero–One Matrices*, Mich. Math. J. **35** (1988), 261–267.

Davis, C.
[1] *Extending the Kantorovich Inequalities to Normal Matrices*, Lin. Alg. Appl. **31** (1980), 173–177.
[2] *Notions Generalizing Convexity for Functions Defined on Spaces of Matrices*, in Convexity (V. Klee, ed.), Amer. Math. Soc., Providence, RI, 1963, 187–201.

de Barra, G.
[1] *The Boundary of the Numerical Range*, Glasgow Math. J. **22** (1981), 69–72.

de Barra, G., Giles, J. R. and Sims, B.
[1] *On the Numerical Range of Compact Operators on Hilbert Spaces*, J. London Math. Soc. **5** (1972), 704–706.

Donoghue, W. F.
[1] *On the Numerical Range of a Bounded Operator*, Mich. Math. J. **4** (1957), 261–263.

Douglas, R. G. and Rosenthal, P.
[1] *A Necessary and Sufficient Condition for Normality*, J. Math. Anal. Appl. **22** (1968), 10–11.

Eiermann, M.
[1] *Fields of Values and Iterative Methods*, Lin. Alg. Appl. **180** (1993), 167–197.

Embry, M.
[1] *The Numerical Range of an Operator*, Pac. J. Math. **32** (1970), 647–650.

Furuta, T.
[1] *On the Class of Paranormal Operators*, Proc. Jpn. Acad. **43** (1967), 594–598.
[2] *Some Characterizations of Convexoid Operators*, Rev. Roum. Math. Pure & Appl. **18** (1973), 893–900.
[3] *Relations between Generalized Growth Conditions and Several Classes of Convexoid Operators,*, Can. J. Math. **29** (1977), 1010–1030.
[4] $A \geq B \geq C$ assures $B^r A^p B^r)^{1/q} \geq B^{p+2r)/q}$ for $r \geq 0$, $p \geq 0$, $q \geq 1$ with $(1+2r)q \geq p+2r$, Proc. Amer. Math. Soc. **101** (1987), 85-88.
[5] *Extension of the Furuta Inequality and Ando–Hiai Log–Majorization*, Lin. Alg. Applic. **219** (1995), 139–155.

Furuta, T., Izumino, S. and Prasanna, S.
[1] *A Characterization of Centroid Operators*, Math. Jpn. **27** (1982), 105–106.

Furuta, T. and Nakamoto, R.
[1] *Certain Numerical Radius Contraction Operators*, Proc. Amer. Math. Soc. **29** (1971), 521–524.

Garske, G.
[1] *The Boundary of the Numerical Range of an Operator*, J. Math. Anal. Appl. **68** (1979), 605–607.

Gersgorin, S.
[1] *Über die Abrenzung der Eigenwerte einer Matrix*, Izv. Akad. Nauk SSSR (Ser. Mat.) **7** (1931), 749–754.

Goldberg, M.
[1] *On Certain Finite Dimensional Numerical Ranges and Numerical Radii*, Lin. Multilin. Alg. **7** (1979), 329–342.

Goldberg, M. and Straus, E. G.
[1] *Elementary Inclusion Relations for Generalized Numerical Ranges*, Lin. Alg. Appl. **18** (1977), 11–24.
[2] *Norm Properties of C-Numerical Radii*, Lin. Alg. Appl. **24** (1979), 113–131.

Goldberg, M. and Tadmor, E.
[1] *On the Numerical Radius and Its Applications*, Lin. Alg. Appl. **42** (1982), 263–284.

Goldberg, M., Tadmor, E. and Zwas, G.
[1] *The Numerical Radius and Spectral Matrices* Lin. Multilin. Alg. **2** (1975), 317–326.

Goldberg, M. and Zwas, G.
[1] *On Matrices Having Equal Spectral Radius and Spectral Norm*, Lin. Alg. Appl. **8** (1974), 427–434.
[2] *Inclusion Relations Between Certain Sets of Matrices*, Lin. Multilin. Alg. **4** (1976), 55–60.

G. H. Golub, G. H. and C. F. Van Loan, C. F.
[1] *Matrix Computations*, Johns Hopkins Univ. Press, Baltimore, MD, 1989.

Goodrich, R. and Gustafson, K.
[1] *Weighted Trigonometric Approximation and Inner–Outer Functions on Higher Dimensional Euclidean Spaces*, J. Approx. Theory **31** (1981), 368–382.
[2] *Spectral Approximation*, J. Approx. Theory **48** (1986), 272–293.

Grone, R., Johnson, C., Sa, E. and Wolkowicz, H.
[1] *Normal Matrices*, Lin. Alg. Appl. **87** (1987), 213–225.

Gustafson, K.
[1] *A Perturbation Lemma*, Bull. Amer. Math. Soc. **72** (1966), 334–338.
[2] *The Angle of an Operator and Positive Operator Products*, Bull. Amer. Math. Soc. **74** (1968), 488–492.
[3] *Positive (Noncommuting) Operator Products and Semigroups*, Math. Z. **105** (1968), 160–172.
[4] *A Note on Left Multiplication of Semigroup Generators*, Pac. J. Math. **24** (1968), 463–465.
[5] *A Min-Max Theorem*, Not. Amer. Math. Soc. **15** (1968), 799.
[6] *Polynomials of Infinitesimal Generators of Contraction Semigroups*, Not. Amer. Math. Soc. **16** (1969), 767.
[7] *The Toeplitz–Hausdorff Theorem for Linear Operators*, Proc. Amer. Math. Soc. **25** (1970), 203–204.
[8] *Necessary and Sufficient Conditions for Weyl's Theorem*, Mich. Math. J. **19** (1972), 71–81.
[9] *Antieigenvalue Inequalities in Operator Theory*, in Inequalities III, Proc. Los Angeles Symposium, 1969 (O. Shisha, ed.), Academic Press, New York, 1972, 115–119.
[10] *The RKNG (Rellich–Kato–Nagy–Gustafson) Perturbation Theorem for Linear Operators in Hilbert and Banach Space*, Acta Sci. Math. **45** (1983), 201–211.

[11] *Partial Differential Equations*, 2nd ed., Wiley, New York, 1987.
[12] *Antieigenvalues in Analysis*, Proc. Fourth International Workshop in Analysis and Its Applications, June 1–10, 1990, Dubrovnik, Yugoslavia (C. Stanojevic and O. Hadzic, eds.), Novi Sad, Yugoslavia, 1991, 57–69.
[13] *Applied Partial Differential Equations I, II*, Kaigai Publishers, Tokyo, Japan, 1991, 1992 (in Japanese).
[14] *Introduction to Partial Differential Equations and Hilbert Space Methods*, 3rd ed., International Journal Services, Calcutta, India, 1993.
[15] *Operator Trigonometry*, Lin. Multilin. Alg. **37** (1994), 139–159.
[16] *Antieigenvalues*, Lin. Alg. Appl. **208/209** (1995), 437–454.
[17] *Matrix Trigonometry*, Lin. Alg. Appl. **217** (1995), 117–140.
[18] *Operator Angles (Gustafson), Matrix Singular Angles (Wielandt), Operator Deviations (Krein)*, Collected Works of Helmut Wielandt, II (B. Huppert and H. Schneider, eds.), De Gruyters, Berlin, 1996.
[19] *Trigonometric Interpretation of Iterative Methods*, Proc. Conf. on Algebraic Multilevel Iteration Methods with Applications, (O. Axelsson, ed.), June 13-15, 1996, Nijmegen, Netherlands, 23–29.
[20] *The Geometrical Meaning of the Kantorovich–Wielandt Inequalities*, (1996), to appear.
[21] *Lectures on Computational Fluid Dynamics, Mathematical Physics, and Linear Algebra*, Kaigai Publications, Tokyo, 1996.
[22] *Operator Trigonometry of Iterative Methods*, Num. Lin. Alg. Applic., (1996), to appear.
[23] *Operator Trigonometry of the Model Problem*, (1996), to appear.
[24] *Operator Spectral States*, Computers Math. Applic., (1996), to appear.

Gustafson, K. and Hartman, R.
 [1] *Graph Theory and Fluid Dynamics*, SIAM J. Alg. Disc. Math. **6** (1985), 643–656,

Gustafson, K. and Lumer. G.
 [1] *Multiplicative Perturbation of Semigroup Generators*, Pac. J. Math. **41** (1972), 731–742.

Gustafson, K. and Rao, D.
 [1] *Numerical Range and Accretivity of Operator Products*, J. Math. Anal. Appl. **60** (1977), 693–702.
 [2] *Spectral States of Normal-like Operators*, (1996), to appear.

Gustafson, K. and Sato, K. I.
 [1] *Some Perturbation Theorems for Nonnegative Contraction Semigroups*, J. Math. Soc. Jpn. **21** (1969), 200–204.

Gustafson, K. and Seddighin, M.
 [1] *Antieigenvalue Bounds*, J. Math. Anal. Appl. **143** (1989), 327–340.

[2] *A Note on Total Antieigenvectors*, J. Math. Anal. Appl. **178** (1993), 603–611.

Gustafson, K. and Zwahlen, B.
[1] *On the Cosine of Unbounded Operators*, Acta Sci. Math. **30** (1969), 33–34.
[2] *On Operator Radii*, Acta Sci. Math. **36** (1974), 63–68.

Haagerup, U. and de la Harpe, P.
[1] *The Numerical Radius of a Nilpotent Operator on a Hilbert Space*, Proc. Amer. Math. Soc. **115** (1992), 371–379.

Hackbusch, W.
[1] *Iterative Solution of Large Sparse Systems of Equations*, Springer-Verlag, Berlin, 1994.

Halmos, P.
[1] *Normal Dilations and Extensions of Operators*, Summa Brasil. Math. **2** (1950), 125–134.
[2] *A Hilbert Space Problem Book*, 2nd ed., Springer-Verlag, New York, 1982.

Hausdorff, F.
[1] *Der Wertvorrat einer Bilinearform*, Math. Z. **3** (1919), 314–316.

Heinz, E.
[1] *Beiträge zur Störungstheorie der Spektralzerlegung*, Math. Ann. **123** (1951), 415–438.

Hess, P.
[1] *A Remark on the Cosine of Linear Operators*, Acta Sci. Math. **32** (1971), 267–269.

Hildebrandt, S.
[1] *The Closure of the Numerical Range as a Spectral Set*, Comm. Pure Appl. Math. (1964), 415–421.
[2] *Über den Numerischen Wertebereich eines Operators*, Math. Ann. **163** (1966), 230–247.

Hille, E. and Phillips, R.
[1] *Functional Analysis and Semigroups*, Amer. Math. Soc. Colloq. Publ. **31** (1957).

Holbrook, J. A. R.
[1] *On the power-bounded operators of Sz.-Nagy and Foias*, Acta Sci. Math. **29** (1968), 299–310.
[2] *Multiplicative Properties of the Numerical Radius in Operator Theory,"* J. Reine Angew. Math. **237** (1969), 166–174.

Horne, R. A.
[1] *The Hadamard Product*, Proc. Symp. in Appl. Math. **40** (1990), 87–120.

Horne, R. A. and Johnson, C. R.
[1] *Matrix Analysis*, Cambridge Univ. Press, Cambridge, 1985.
[2] *Topics in Matrix Analysis*, Cambridge Univ. Press, Cambridge, 1991.

Hughes, G.
[1] *A Note on the Shape of the Generalized Numerical Range*, Lin. Multilin. Alg. **26** (1990), 43–47.

Istratescu, V. and Istratescu, I.
[1] *On Normaloid Operators*, Math. Z. **105** (1967), 153–156.
[2] *On Bare and Semibare Points for Some Classes of Operators*, Port. Math. **29** (1970), 205–211.

Johnson, C. R.
[1] *A Gersgorin inclusion set for the field of values of a finite matrix*, Proc. Amer. Math. Soc. **41** (1973), 57–60.
[2] *Gershgorin Sets and the Field of Values*, J. Math. Anal. Appl. **45** (1974), 416–419.
[3] *Normality and the Numerical Range*, Lin. Alg. Appl. **15** (1976), 89–94.
[4] *Numerical Determination of the Field of Values of a General Complex Matrix*, SIAM J. Numer. Anal. **15** (1978), 595–602.

Johnson, C. R., Spitkovsky, I. M. and Gottlieb, S.
[1] *Inequalities Involving the Numerical Radius*, Lin. Multilin. Alg. **37** (1994), 13–24.

Jones, M. S.
[1] *A Note on the Shape of the Generalized C-Numerical Range*, Lin. Multilin. Alg. **31** (1991), 81–84.

Kato, T.
[1] *Some Mapping Theorems for the Numerical Range*, Proc. Jpn. Acad. **41** (1965), 652–655.
[2] *Perturbation Theory for Linear Operators*, 2nd ed., Springer-Verlag, New York, 1980.

Kern, M., Nagel, R., and Palm, G.
[1] *Dilations of Positive Operators: Construction and Ergodic Theory*, Math. Z. **156** (1977), 265–267.

Krein, M.
[1] *Angular Localization of the Spectrum of a Multiplicative Integral in a Hilbert Space*, Func. Anal. Appl. **3** (1969), 89–90.

Kreiss, H. O.
[1] *Über die Stabilitätsdefinition für Differenzengleichungen die partielle Differentialgleichungen approximieren*, BIT **2** (1962), 153–181.

Kumar, R. and Das, P.
[1] *Perturbation of m-Accretive Operators in Banach Spaces*, Nonlin. Anal. **17** (1991), 161–168.

Kyle, J.
[1] W_δ *is Convex*, Pac. J. Math. **72** (1977), 483–485.

Lax, P. and Phillips, R. S.
[1] *Scattering Theory*, Academic Press, New York, 1967.

Lax, P. and Wendroff, B.
[1] *Systems of Conservation Laws*, Comm. Pure Appl. Math. **13** (1960), 217–237.
[2] *Difference Schemes for Hyperbolic Equations with High Order of Accuracy*, Comm. Pure Appl. Math. **17** (1964), 381–398.

Le Veque, R. J. and Trefethen, L. N.
[1] *On the Resolvent Condition in the Kreiss Matrix Theorem*, BIT **24** (1984), 585–591.

Lebow, A.
[1] *On Von Neumann's Theory of Spectral Sets*, J. Math. Anal. Appl. **7** (1963), 64–90.

Lenferink, H. W. J. and Spijker, M. N.
[1] *A Generalization of the Numerical Range of a Matrixa*, Lin. Alg. Appl. **140** (1990), 251–266.

Li, C. K.
[1] *C-Numerical Ranges and C-Numerical Radii*, Lin. Multilin. Alg. **37** (1994), 51–82.
[2] *Numerical Ranges of the Powers of an Operator*, preprint, 1995.

Löwner, K.
[1] *Über Monotone Matrixfunktionen*, Math Z. **38** (1934), 177-216.

Luecke, G. R.
[1] *A Class of Operators on Hilbert Space*, Pac. J. Math. **41** (1971), 153–156.

Luenberger, D.
[1] *Linear and Nonlinear Programming*, 2nd ed., Addison–Wesley, Reading, MA, 1984.

Lumer, G.
[1] *Semi-inner-product Spaces*, Trans. Amer. Math. Soc. **100** (1961), 24–43.

Lumer, G. and Phillips, R.
[1] *Dissipative Operators in a Banach Space*, Pac. J. Math. **11** (1961), 679–698.

Macluer, C. R.
[1] *On Extreme Points of the Numerical Range of Normal Operators*, Proc. Amer. Math. Soc. **16** (1965), 1183–1184.

Man, W.
[1] *The Convexity of the Generalized Numerical Range*, Lin. Multilin. Alg. **20** (1987), 229–245.

Marcus, M.
[1] *Basic Theorems in Matrix Theory*, Natl. Bur. Standards Appl. Math., Sec. 57, 1960.
[2] *Some Combinatorial Aspects of the Generalized Numerical Range*, Ann. N. Y. Acad. Sci. **319** (1979), 368–376.
[3] *Computer Generated Numerical Ranges and Some Resulting Theorems*, Lin. Multilin. Alg. **20** (1987), 121–157.
[4] *Matrices and Matlab*, Prentice Hall, Englewood Cliffs, NJ, 1993.

Marcus, M. and Shure, B. N.
[1] *The Numerical Range of Certain 0, 1-Matrices*, Lin. Multilin. Alg. **7** (1979), 111–120.

Mees, A. I. and Atherton, D. P.
[1] *Domains Containing the Field of Values of a Matrix*, Lin. Alg. Appl. **26** (1979), 289–296.

Meng, C. H.
[1] *A Condition That a Normal Operator Has a Closed Numerical Range*, Proc. Amer. Math. Soc. **8** (1957), 85–88.
[2] *On the Numerical Range of an Operator*, Proc. Amer. Math. Soc. **14** (1963), 167–171.

Mirman, B. A.
[1] *Numerical Range and Norm of a Linear Operator*, Trudy Seminara Funk. Anal. **10** (1968), 51–55.
[2] *Antieigenvalues: Method of Estimation and Calculation*, Lin. Alg. Appl. **49** (1983), 247–255.

Miyadera, I.
[1] *Nonlinear Semigroups*, Amer. Math. Soc., Providence, RI, 1992.

Moyls, B. N. and Marcus, M. D.
[1] *Field Convexity of a Square Matrix*, Proc. Amer. Math. Soc. **6** (1955), 981–983.

Müller, V.
[1] *The Numerical Radius of a Commuting Product*, Mich. Math. J. **35** (1988), 255–260.

Murnaghan, F. D.
[1] *On the Field of Values of a Square Matrix*, Proc. Natl. Acad. Sci. U.S.A. **18** (1932), 246–248.

Nirschl, N. and Schneider, H.
[1] *The Bauer Field of Values of a Matrix*, Numer. Math. **6** (1964), 355–365.

Okubo, K. and Ando, T.
[1] *Operator Radii of Commuting Products*, Proc. Amer. Math. Soc. **56** (1976), 203–210.

Orland, G. H.
[1] *On a Class of Operators*, Proc. Amer. Math. Soc. **15** (1964), 75–79.

Parter, S.
[1] *Stability, Convergence, and Pseudo-Stability of Finite-Difference Equations for an Over-Determined Problem*, Numer. Math. **4** (1962), 277–292.

Pearcy, C.
[1] *An Elementary Proof of the Power Inequality for the Numerical Radius*, Mich. Math. J. **13** (1966), 289–291.

Poon, Y.
[1] *Another Proof of a Result of Westwick*, Lin. Multilin. Alg. **9** (1980), 35–37.
[2] *The Generalized k-Numerical Range*, Lin. Multilin. Alg. **9** (1980), 181–186.

Prasanna, S.
[1] *The Norm of a Derivation and the Björck–Thomee–Istratescu Theorem*, Math Jpn. **26** (1981), 585–588.

Radjabalipour, M. and Radjavi, H.
[1] *On the Geometry of Numerical Range*, Pac. J. Math. **61** (1975), 507–511.

Rao, D.
[1] *Numerical Range and Positivity of Operator Products*, Ph.D. thesis, University of Colorado, Boulder, 1972.
[2] *A Triangle Inequality for Angles in a Hilbert Space*, Rev. Colom. Mat. **10** (1976), 95–97.
[3] *Operadores Paranormalesa*, Rev. Colom. Mat. **21** (1987), 135–149.
[4] *Rango Numerico de Operadores Conmutativos*, Rev. Colom. Mat. **27** (1994), 231–233.

Reddy, S. and Trefethen, L. N.
[1] *Stability and the Method of Lines*, Numer. Math. **62** (1992), 235–267.

Richtmyer, R. D. and Morton, K.
[1] *Difference Methods for Initial Value Problems*, Wiley, New York, 1967.

Ridge, W. C.
[1] *Numerical Range of a Weighted Shift with Periodic Weights*, Proc. Amer. Math. Soc. **55** 107–110.

Riesz, F. and Sz. Nagy, B.
[1] *Functional Analysis*, F. Ungar Publishing Co., New York, 1955.

Rokhlin, V.
[1] *Exact Endomorphisms of a Lebesgue Space*, Amer. Math. Soc. Trans. **39** (1964), 1–36.

Saito, T. and Yoshino, T.
[1] *On a Conjecture of Berberian*, Tohoku Math. J. **17** (1965), 147–149.

Saunders, B. D. and Schneider, H.
[1] *A Symmetric Numerical Range for Matrices*, Numer. Math. **26** (1976), 99–105.

Schäffer, J. J.
[1] *On Unitary Dilations of Contractions*, Proc. Amer. Math. Soc. **6** (1955), 322.

Schmid, P. J., Henningson, D. S., Khorrami, M., and Malik, M. R.
[1] *A Study of Eigenvalue Sensitivity for Hydrodynamic Stability Operators*, Theoret. Comput. Fluid Dynamics **4** (1993), 227–240.

Schreiber, M.
[1] *Numerical Range and Spectral Sets*, Mich. Math. J. **10** (1963), 283–288.

Schur, J.
[1] *Bemerkungen zur Theorie der Beschränkten Bilinearformen mit Unend- lich vielen Veränderlichen*, J. Reine Angew. Math. **140** (1911), 1–28.

Shiu, E. S. W.
[1] *Growth of the Numerical Ranges of Powers of Hilbert Space Operators*, Mich. Math. J. **23** (1976), 155–160.

Sims, B.
[1] *On the Connection Between the Numerical Range and Spectrum of an Operator in a Hilbert Space*, J. London Math. Soc. **8** (1972), 57–59.

Solov'ev, V. N.
[1] *A Generalization of Gersgorin's Theorem*, Izv. Akad. Nauk. SSSR Ser. Mat. **47** (1983), 1285–1302; English translation in Math. USSR Izv. **23** (1984).

Spijker, M. N.
[1] *On a Conjecture by Le Veque and Trefethen Related to the Kreiss Matrix Theorem*, BIT **31** (1991), 551–555.
[2] *Numerical Ranges and Stability Estimates*, Appl Num. Math. **13** (1993), 241–249.

Stampfli, J. G.
[1] *Hyponormal Operators*, Pac. J. Math. **12** (1962), 1453–1458.
[2] *Hyponormal Operators and Spectral Density*, Trans. Amer. Math. Soc. **117** (1965), 469–476.
[3] *Extreme Points of the Numerical Range of a Hyponormal Operator*, Mich. Math. J. **13** (1966), 87–89.
[4] *The Norm of a Derivation*, Pac. J. Math. **33** (1970), 737–747.

Starke, G.
[1] *Fields of Values and the ADI Method for Non-normal Matrices*, Lin. Alg. Appl. **180** (1993), 199–218.

Stewart, G. and Sun, J. G.
[1] *Matrix Perturbation Theory*, Academic Press, Boston, 1990.

Stone, M. H.
[1] *Linear Transformations in Hilbert Space*, American Mathematical Society, Providence, RI, 1932.

Stout, Q. F.
[1] *The Numerical Range of a Weighted Shift*, Proc. Amer. Math. Soc. **88** (1983), 495–502.

Sz. Nagy, B.
[1] *Sur les Contractions de l'espace de Hilbert*, Acta Sci. Math. **15** (1953), 87–92.

Sz. Nagy, B. and Foias, C.
[1] *On Certain Classes of power-bounded operators in Hilbert space*, Acta. Sci. Mat. **27** (1966), 17–25.
[2] *Harmonic Analysis of Operators on Hilbert Space*, North-Holland, Amsterdam, 1970.

Toeplitz, O.
[1] *Das algebraische Analogon zu einem Satz von Fejér*, Math. Z. **2** (1918), 187–197.

Trefethen, L. N.
[1] *Approximation Theory and Numerical Linear Algebra*, in Algorithms for Approximation II (J. Mason and M. Cox, eds.), Chapman, London, 1990, 336–360.

Trefethen, L. N., Trefethen, A. E. and Reddy, S., Driscoll, T.
[1] *Hydrodynamic Stability without Eigenvalues*, Science **261** (1993), 578–584.

Tsing, N. K.
[1] *On the Shape of the Generalized Numerical Range*, Lin. Multilin. Alg. **10** (1981), 173–182.

Tsing, N. K. and Cheung, W. S.
[1] *Star-Shapedness of the Generalized Numerical Ranges*, in Abstracts 3rd Workshop on Numerical Ranges and Numerical Radii (T. Ando and K. Okubo, eds.), Sapporo, Japan, 1966.

Varah, J. M.
[1] *On the Separation of Two Matrices*, SIAM J. Numer. Anal. **16** (1979), 216–222.

Von Neumann, J.
[1] *Eine Spektraltheorie für Allgemeine Operatoren eines Unitären Raumes*, Math. Nachr. **4** (1951), 258–481.

Westwick, R.
[1] *A Theorem on Numerical Range*, Lin. Multilin. Alg. **2** (1975), 311–315.

Wielandt, H.
[1] *National Bureau of Standards Report* 1367, NBS, Washington, DC, 1951.
[2] *Topics in the Analytic Theory of Matrices*, Univ. of Wisconsin Lecture Notes, Madison, 1967.

Williams, J. P.
[1] *Spectra of Products and Numerical Ranges*, J. Math. Anal. Appl. **17** (1967), 214–220.

Williams, J. P. and Crimmins, T.
[1] *On the Numerical Radius of a Linear Operator*, Amer. Math. Mon. **74** (1974), 832–833.

Wintner, A.
[1] *Zur Theorie der beschrankten Bilinearformen*, Math. Z **30** (1929), 228–282.

Yosida, K.
[1] *Functional Analysis*, Springer-Verlag, Berlin, 1966.

Young, D. M.
[1] *Iterative Solution of Large Linear Systems*, Academic Press, New York, 1971.

Zarantonello, E.
[1] *The Closure of the Numerical Range Contains the Spectrum*, Pac. J. Math. **22** (1967), 575–595.

Index

Accretive, 25, 49, 52, 61, 71
Algebraic multiplicity, 124, 163
Algorithm, optimization, 81, 84, 87, 108
Analytical functions, 29, 31, 43
Angle of an operator, 49, 83, 86
Anti-eigenvalue, functional, 75, 77
Anti-eigenvalues, 67, 74, 79, 86, 168
Anti-eigenvalues, total, 67, 69
Antieigenvectors, 67, 73, 74

Banach–Lax theorem, 89
Bare points, 151

Chain, 119
Condition number, 82
Consistency condition, 89
Convexity, 3, 73
Corner, 19
Cosine of an operator, 49
Cosine, total, 49
Cramped spectrum, 35
Cycle, 119

Decomposition, 110
Deviation, 56, 60
Dilation, 42, 45
Dilation theory, 21, 42, 45
Dilation, ρ-, 43
Dilation, normal, 160
Discrete boundedness, 91

Discrete stability, 88
Doubly stochastic, 131

Eigenvalue, 3, 9, 68
Eigenvalue, normal, 113, 163
Ellipse lemma, 3
Envelope, 5
Euler equation, 75
Euler explicit scheme, 88, 93
Euler system, 96
Extreme point, 16, 18, 111

Field of values, 1
Finite difference scheme, 88, 93
Fluid dynamics, 93, 105
Friedrich's extension, 65

Galerkin System, 95
Gas dynamics, 96, 99, 106
Gersgorin sets, 113, 133
Grammian, 57
Graph radius, 120

Hadamard product, 123
Hahn–Banach theorem, 22
Heat equation, 88, 91
Herglotz's theorem, 31
Hessian, 82

Implicit schemes, 95
Infinitesimal generator, 59, 95
Initial value problem, 60, 88
Isometry, 38

Jacobian, 97

Kantorovich bound, 81
Kantorovich inequality, 50
Kernel, 32, 169
Kronecker product, 124

Lagrange function, 8
Lax equivalence theorem, 89
Lax–Wendroff scheme, 98
Line of support, 19, 33
Linear solver, 108, 149

Matrix norm, 128
Matrix, 0–1, 118
Matrix, companion, 117, 122
Matrix, diffusion, 95
Matrix, normal, 112
Min-max equality, 53

Navier–Stokes equations, 93
Numerical analysis, 80
Numerical boundary, 18
Numerical radius, 8, 24, 28, 37, 109 115
Numerical radius, C, 128
Numerical range, 1, 22, 25
Numerical range, δ-, 134
Numerical range, algebraic, 25, 92, 133
Numerical range, augmented, 103
Numerical range, C, 22, 127
Numerical range computation, 137
Numerical range, F, 132
Numerical range, generalized, 127
Numerical range, k, 22, 132
Numerical range, M, 22, 92, 133
Numerical range, restricted, 22, 134
Numerical range, sesquilinear, 22
Numerical range, spatial, 23, 132
Numerical range, symmetric, 22, 135

Operator products, 34

Operator trigonometry, 49, 107
Operator, accretive, 25, 49, 52, 61
Operator, centroid, 159
Operator, convexoid, 150, 154, 155, 163
Operator, dissipative, 52, 60, 152
Operator, hyponormal, 156
Operator, idempotent, 13
Operator, normal, 15, 39, 69, 71, 162
Operator, normaloid, 150, 153, 159, 163
Operator, paranormal, 156
Operator, sectorial, 64, 66
Operator, selfadjoint, 7, 15, 36, 62, 66, 85
Operator, shift, 2, 41, 45, 120, 167
Operator, spectraloid, 150, 155, 163, 166
Operator, subnormal, 166
Operator. transloid, 163
Operator, unbounded, 24, 51
Operator, unitary, 35, 43, 110

Perturbation, additive, 59
Perturbation, multiplicative, 59
Pinching, 130
Positively stable, 35
Power boundedness, 90
Power dilation, 29
Power equality, 155
Power inequality, 28, 37, 99
Pseudo eigenvalues, 103, 107

Rational functions, 32, 159
Resolvent, 8
Resolvent growth, 17, 150, 152, 168
RKNG theorem, 59

Schur's lemma, 164
Schur decomposition, 4
Schur triangularization, 110
Semigroup, contraction, 59
Semigroup generators, 47, 58, 60

Sine of an operator, 53
Singular angle, 58
Semi-inner product, 22
Spectral inclusion, 6, 23
Spectral radius, 10, 166
Spectral set, 159, 161
Spectrum, 6, 10, 24
Submatrix inclusion, 42, 110

Test function, 95

Toeplitz–Hausdorff theorem, 4
Transcendental radius, 159

Uniform Boundedness principle, 90

Value field, 109
Von Neumann–Kreiss stability, 90, 92

Universitext *(continued)*

Rubel/Colliander: Entire and Meromorphic Functions
Sagan: Space-Filling Curves
Samelson: Notes on Lie Algebras
Schiff: Normal Families
Shapiro: Composition Operators and Classical Function Theory
Simonnet: Measures and Probability
Smith: Power Series From a Computational Point of View
Smoryński: Self-Reference and Modal Logic
Stillwell: Geometry of Surfaces
Stroock: An Introduction to the Theory of Large Deviations
Sunder: An Invitation to von Neumann Algebras
Tondeur: Foliations on Riemannian Manifolds
Zong: Strange Phenomena in Convex and Discrete Geometry

**QA 329.2 .G88 1997
Gustafson, Karl E.
Numerical range**